John Wood

Lectures on Hernia and its Radical Cure

Delivered at the Royal College of Surgeons of England in June, 1885 - with

A Clinical Lecture on Trusses and Their Application to Ruptures

John Wood

Lectures on Hernia and its Radical Cure
Delivered at the Royal College of Surgeons of England in June, 1885 - with A Clinical Lecture on Trusses and Their Application to Ruptures

ISBN/EAN: 9783337162047

Printed in Europe, USA, Canada, Australia, Japan

Cover: Foto ©berggeist007 / pixelio.de

More available books at **www.hansebooks.com**

LECTURES

ON

HERNIA AND ITS RADICAL CURE.

LECTURES

ON

HERNIA AND ITS RADICAL CURE.

*Delivered at the Royal College of Surgeons of
England, in June, 1885.*

WITH

A CLINICAL LECTURE ON TRUSSES AND THEIR APPLICATION TO RUPTURES,

Delivered at King's College Hospital.

BY

JOHN WOOD, F.R.S., F.R.C.S.

VICE-PRESIDENT AND HUNTERIAN PROFESSOR OF SURGERY AND PATHOLOGY IN
THE ROYAL COLLEGE OF SURGEONS, SENIOR SURGEON AND PROFESSOR
OF CLINICAL SURGERY TO KING'S COLLEGE HOSPITAL.

HENRY RENSHAW,
356 STRAND, LONDON.
1886.

PREFACE.

THIS volume is published, at the suggestion of professional friends, as a more convenient form for reference than that afforded by the various Medical Journals, in which the Lectures have already appeared.

It is reprinted from the *British Medical Journal*, with the addition of original woodcuts, necessary for the illustration and better explanation of a subject difficult to elucidate well, and at the same time compendiously, by the unaided text.

A Clinical Lecture on the application of trusses from the same source has been added to render the work more complete.

And I take the opportunity to express, by way of dedication, my sense of gratitude to the President and Council of the Royal College of Surgeons of England, for enabling me to make known the results of my experience, and such advance as I have been able to make in this department of surgery, from the chair of the Hunterian Professor of Surgery and Pathology to the College.

61, WIMPOLE STREET,
     CAVENDISH SQUARE.
     November, 1885.

CONTENTS.

—:—

LECTURE I.

LECTURE II.

LECTURE III.

HERNIA AND ITS RADICAL CURE.

LECTURE I.

Introduction.—Limits of the subject.—Inguinal, Crural, and Umbilical Hernia.—Visceral and Parietal Causes.—Opinions acted on by modern Surgeons.—Structure and Development of the Hernial Areas.—Descent of Testis.—Imperfections of Evolution as Causes of Hernia.—Anatomy of the Parts of Inguinal Hernia.—Structure and Mechanism of the Canal and Rings in Health and Disease.— Oblique and Direct Hernia ; Congenital, Infantile.—Relation to forms of Hydrocele.

MR. PRESIDENT AND GENTLEMEN, — In undertaking the duty with which I have had the honour to be entrusted by the Council of this College, my selection of the subject of my lectures was naturally influenced by the fact that, twenty-four years ago, the authorities of this College, in conferring upon me the Jacksonian prize for an essay embodying the results of a new method of operating for the radical cure in sixty cases of inguinal hernia, encouraged me greatly in the investigation of this field of research.

Fully and deeply impressed with the idea that it was only by patient and long-sustained efforts that a subject of so much difficulty in obtaining and recording true results, for periods long enough to be worthy of reliance, could be satisfactorily elucidated, I have, during this long interval, continued to modify,

B

to perfect, and to extend the scope of the method of operating which I then published for the first time.

The attention and efforts of surgeons have been turned in this direction from the earliest ages of surgical art. New methods have continually been brought forward, and old methods resuscitated, wholly or in part, with various degrees of success or failure. In no subject has the plausibility of theory been more contrasted with futility of result. The delimitation between the safety of success and the fatality of failure has proved most difficult to determine and to mark out.

Yet notwithstanding, I think I may now say, with a due regard to scientific exactitude, that operative treatment, with a view to produce a radical cure of hernia, has gained ground in the confidence of the medical profession; and that the disappointing results and repeated failures of much-lauded methods have not entirely established amongst us that despairing condition of opinion of which Lord Bacon says, "that he will not doubt to note as a deficience," in the practitioners of his time, "that they inquire not the perfect cure of many diseases, but by pronouncing them incurable, do enact a law of neglect, and exempt ignorance from discredit" ("Advancment of Learning," book ii. p. 43).

In an age when all things are tested and brought to proof again and again, no exception can be made in a matter which so seriously affects the welfare of the very large number of our fellow-creatures who suffer and die from hernia.

In the course of the three lectures which are allotted to me I shall not have more time than barely to deal with my own work in the wide field of hernia.

A subject absolutely overloaded with literature and bibliography will not fairly permit of scant notices and mere honourable mentions. To read out the names and to summarise the work of the labourers in this field only, would probably occupy the whole of the three lectures, and then would only imperfectly do the work of the many excellent text-books and encyclopædias. I shall venture, however, further on, to classify very briefly the varieties of the operative treatment of hernia. I shall confine myself to the more common forms of hernia, the most troublesome and the most disabling, and therefore the more urgently calling for a more complete and satisfactory treatment than constant and usually imperfect truss pressure—namely, inguinal, crural, and umbilical hernia.

During the twenty-seven years which have elapsed since my first operation for the radical cure of hernia I have operated upon 391 cases; namely, 370 inguinal, 16 crural, and 5 umbilical.

In 11 cases the hernia was double, and both sides were operated on; in 12 cases, operations, which at first failed, were repeated (in two cases three times); making a grand total of 414 operations.

Of the 370 cases of inguinal hernia, 219 were found on the right side, 128 on the left side, and 23 on both sides; 10 of them occurred in females; 123 were cases of congenital origin, occurring either in children or seen in patients above fourteen years of age, with a history of congenital origin; 114 occurred in males, and 9 in females; of the males, 51 were found on the right side, 41 on the left side, and 13 on both sides. In 9 male cases one of the testes was undescended, and placed at or above the

superficial ring.   Six of the undescended testes
were on the right side and 3 on the left.   Of the
9 females, 6 occurred on the right side and 3 on
the left.

Of the 15 cases of crural hernia, 10 were found
on the right side and 5 on the left; 3 were in
males and 12 in females.

The 5 cases of umbilical hernia were all males.

Pathology and Causes of Inguinal Hernia.
—In estimating these, we have to consider—first,
the contained viscera of the peritoneal cavity, and
their action under the influence of the muscles which
surround and compress them ; and secondly, the
formation of the containing walls, and the reason of
their yielding in certain hernial regions.   Of the
contained viscera, the most likely to become displaced
are the omentum and small intestines, the former
because of its variable length and the mobility of
its attachments to the stomach and colon ; and the
latter because of the extensibility of the mesentery,
necessitated by the alternating distension and contraction of the muscular tube which it encloses and
retains.   The colon, although frequently implicated
in hernial tumours, especially in the umbilical variety,
is less liable, in its ascending and descending
portions, to hernial displacement, because of the
shortness of the peritoneal processes which attach
them to the abdominal walls in the shape of meso-
colon.   The more solid viscera, both abdominal
and pelvic, although sometimes implicated in hernial
protrusions, when large and of long duration, seem
to be always dragged down either by adhesions
or by their connections with the looser viscera or
peritoneum.

It was held by Richter, Morgagni, and others of the older school of modern surgery, that the immediate and most powerful cause of hernia, especially inguinal, was an elongation of the mesentery, allowing of such an extent of play to the movements of the bowels as to permit protrusion through the weaker parts of the containing walls, and that a mesentery of normal length would not allow of such protrusion. Amongst the many able and experienced of their contemporaries Scarpa stands eminent for the ability with which he contended against this theory, and maintained that the elongation of the mesentery does not, in the formation of a hernia, precede the displacement of the intestine, but is simultaneous with it, and caused by the dragging of the already displaced bowel.

And there are many facts which appear to favour this opinion, and the conclusions of Samuel Cooper, in support of the same view of the question.

Distension of the bowels by food or air, and the muscular action upon them, presses the anterior wall of the abdomen forwards, by a simultaneous and equally distributed force in its action and reaction, being equal and similar almost to that of fluid pressure. The same distension tends, by separating the folds of the mesentery, to shorten its hold upon the bowel, and so to prevent its protrusion through the abdominal wall. When the abdominal cavity is opened by a wound, no protrusion occurs unless and until the bowels are distended with food, fluids, or air, and then it occurs freely ; and it is common to find in the *post-mortem* room a considerable part of the intestines lying in the cavity of the pelvis. If the bowels lie so far down as to rest on the pelvic viscera, which they commonly and normally do, they may certainly reach as far as the groin or deep inguinal and crural hernial apertures, which lie on a superior level. In expiration they are pressed still nearer to the spinal column, close against which, in thin persons, the muscular wall of the abdomen lies. In the normal condition the mesentery of the jejunum, placed towards the left side, is the longest and loosest portion. At a point between the distance of 6 feet to 11 feet from the duodenum, Mr. Treves has shown that the mesentery attains to from 9 to 10 inches in length, from its root to its attachment to the bowel, the lower part measuring from 8 to 9 inches only. ("Hunterian Lectures on the Anatomy of the Intestinal Canal and Peritoneum.")

At the right iliac fossa the lower part of the mesentery measures only about one-half of this. The same observer calls attention to the fact that just after birth the development of the upper part of the small intestine is much more rapid than that of the lower or that of the colon, which is comparatively late in its evolution. This depends upon the physiological changes resulting from the first imbibition of food through the mouth, in the form calculated for easy and rapid and more or less complete absorption, with but little residuum in the shape of fæces. These by their presence stimulate the growth and action of the large intestine, heretofore a simple receptacle for the meconium. The cæcum also is late in reaching to its final resting-place in the right iliac fossa.

The early position of the stomach on the left side, with a more vertical direction than in the adult, tends to place the omentum more on the left side, where in the fœtus and infants it is chiefly aggregated. This is apparently the cause why, in early life, omentum is more frequently and more abundantly found in an inguinal hernia on the left side than on the right. The later descent of the cæcum and ileum on the right side seems to be more than compensated for by the ultimate downward development of the mesentery and intestines on the right side, inasmuch as inguinal hernia is found much more frequently on the right side than on the left. In my own cases, as before stated numerically, inguinal hernia was found almost twice as frequently on the right side as on the left. The chief reason of this will be found to lie in causes referring rather to the effect of the evolution of the testicle upon the abdominal wall. In about the same proportion of two to one, this gland was found to be retained in the abdomen or groin more frequently on the right side than on the left.

Again, we do not uncommonly find in the dissecting-room very long and lax and fatty polypoidal developments of the mesentery, and a very low attachment of its root to the hinder wall of the abdomen, with the whole of the contents of the lower abdominal regions pushed down as it were by a large liver or a hypertrophied stomach. And yet in such cases it is rare to find a hernia coexistent.

In one case out of a hundred examined by Mr. Treves the intestines, in an old woman of seventy, reached as far down out of the belly as eight inches below the anterior superior iliac spine, and yet there was no hernia of any kind. I have myself found many similar cases in my dissecting-room experience.

On the other hand, in normal instances, as Mr. Birkett has pointed out and Mr. Treves has verified, the intestines cannot be artificially drawn down below the level of the pubic spine, either into the scrotum or through the crural canal. It is evident, therefore, that in hernial cases, after the bowel or omentum has become engaged in the deep hernial rings, as far as which they normally reach, the omentum and mesentery, or both, must be elongated by a constant dragging pull upon these structures. And in point of fact nothing is more common than to find in hernial cases a long, pointed process of the omentum, more fatty and thicker than the rest, lying in the sac; or a coil of intestine which had been habitually the inmate of a hernial sac, projecting prominently beyond its neighbouring coils, with a longer mesentery than the rest.

The bearing of these considerations upon a proper judgment of the subject of the radical cure of hernia is evident. If the omentum is elongated and so formed as to slip easily and continually into a hernial sac, and keeping it open, or pressing upon the deeper hernial openings, an operation for the radical cure cannot really be sufficiently effective, unless a portion of the omentum be also removed at the same time. Such removal of the omentum I have, in my later operations, always effected when required. Any attempt at shortening of the mesentery seems so far, even in these days of daring abdominal enterprises, to be out of the scope of practical surgery.

The conclusion we may draw from the foregoing observations is, that while the condition of disproportionate development of the intestines and abdominal viscera may predispose to and assist in the formation of a rupture, the chief causes lie in the imperfection of the structures from delayed evolution in and about the deep hernial apertures. As to the precise determination of the structures in which this deficiency of retaining power resides, there is again much difference of opinion. Before discussing this it will be necessary for us to consider the arrangement of the structures composing the abdominal wall, and their relation to the formation of hernia.

Anatomy of the Parts of Inguinal Hernia. —The more persisting and continuous support is evidently afforded by the strong aponeurotic and tendinous structures, which are thickest and strongest where resistance is most required—viz., in the groins.

Under the subcutaneous and tegumentary layers comes the aponeurotic tendon of the external oblique, passing in parallel fibres obliquely downwards, forwards, and inwards, to be connected at the linea alba with the similar ones on the opposite side, and blended more or less with the tendons of the internal oblique and transversalis muscles lying beneath them. The muscular fibres of the external oblique terminate in an elliptical border at the level of the iliac spine. From this point an intimate combination of the outermost of the tendinous fibres with the fascia lata of the thigh, under the name of Poupart's ligament, passes across the deeper groin structures, to be implanted upon the pubic spine and pectineal line. The iliac portion of fascia lata is connected with it below, from one point to the other, continuously, and holds it down in such a way that the superficial surface is concave, and forms the hollow of the groin, for the accommodation of the lymphatics and superficial vessels.

At the pectineal line the deeper fibres are blended with and help to form a triangular offset, with a lunated border outwards (Gimbernat's ligament), and

FIG. I.

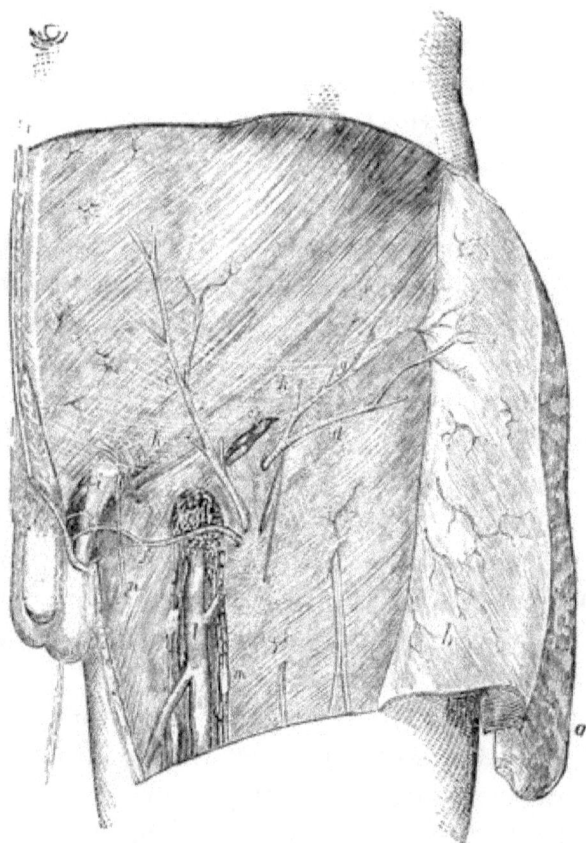

Superficial Dissection of the Inguinal Region : showing superficial inguinal, and saphenous openings.

here the fibres meet, and are united to the pubic portion of fascia lata, to the conjoined tendon, the triangular fascia, and to the deeper fascia, namely, the fascia iliaca and transversalis, which strongly bind it down, and keep it in its curved shape. These important parts are in relation therefore both to inguinal and crural hernia.

Above Poupart's ligament the tendon of the external oblique is covered by the deep, loose, and movable layer of superficial fascia (Scarpa's fascia), which, giving off a wide elastic process of attachment to it, is itself continuous with the deep layer of superficial fascia of the thigh, and contains between its layers the lymphatic glands, vessels, and nerves before mentioned.

Of these vessels, the only ones that need be considered in relation to our subject are the superficial epigastric and the superficial and deep external pubic branches of the common femoral. Of these, the first-named emerge through the saphenous opening, or near it, pass upwards and inwards over Poupart's ligament and the deep abdominal ring, indicating the course, tolerably closely, of the deep vessels of the same name which lie on the peritoneum.

The superficial external pubic emerges also through the saphenous opening, and crosses upwards and inwards over the superficial abdominal ring towards the *mons veneris*, while the deeper ones of the same name emerge through the fascia lata, cross the pubis behind the spermatic cord, to anastomose on the scrotum and penis with other arteries.

The cutaneous nerves in relation with these are the ilio-inguinal, emerging from the superficial ring, to be distributed to the scrotum and lower groin on the thigh. Other small branches of the lower dorsal cutaneous nerves pierce the abdominal aponeurosis at varying distances from the linea alba, with similar

branches of the lower intercostal arteries, to anastomose with the superficial epigastric. The genital
branch of the genito-crural nerve also emerges upon
the spermatic cord or round ligament, through the
superficial abdominal ring, to become distributed, in
the male, to the cremaster muscle and dartos scroti
in the scrotum ; while the crural branch of the same
nerve pierces the fascia lata below Poupart's ligament,
to be distributed on the upper and front part of the
thigh, anastomosing with the internal cutaneous
branch of the anterior crural.

Above Poupart's ligament the aponeurosis of the
external oblique forms an · aperture for the trans-
mission of the spermatic cord or round ligament, the
superficial (external) abdominal ring. The sides of
this opening form, with the pubic crest, an elongated oblique triangle, of which the outer (the external pillar of the ring) lies also inferior, and blends
or is identical with the lower and inner half of
Poupart's ligament.

Flat at its outer half, .it becomes rounded and
cord-like at the inner or lower part, where it is implanted firmly upon the pubic spine. Outside this
insertion it is grooved obliquely for the passage and
lodgment of the spermatic cord, which, on emerging
from the ring, lies outside the pubic spine, which
forms an important internal relation to it, useful in
protecting it from injury, and in diagnosis. This
groove is found to be continuous with one formed in
the canal by the attachment of the fascia transversalis and conjoined tendon to the deep surface of
Poupart's ligament, in which the cord rests on its
way through the inguinal canal, and which forms the
lower boundary of that canal.

The internal pillar of the superficial ring is flat and riband-like, and can be traced downwards to the front of the pubic bone and symphysial ligaments below the pubic crest, forming in the male a part of the ligamentum suspensorium penis. The triangular opening left between the pillars of the superficial ring is converted into an oval opening by the arrangement of some curved cross fibres, in bundles of an inverted arch-shape (arciform fibres), closely connected by a thin but dense fascia, and deeply adherent to the tendinous pillars. The arciform fibres arise in a single bundle from the anterior superior iliac spine. They spread out, to be lost in the aponeurotic fascia, over the lower part of the linea alba. Below, the arciform fascia is continued as a sleeve-like investment over the spermatic cord, forming the intercolumnar or external spermatic fascia, which passes down into the scrotum upon the tunica vaginalis, blending with the cremaster fascia, which lies next below it. When the arciform fascia is divided, the pillars of the ring separate under pressure from within, proving that this fascia forms a barrier against dilatation or separation of the pillars of the ring in the formation of a hernia.

Next in order, under the external oblique aponeurosis, come the internal oblique and transversalis muscles, with the cremasteric fibres all covered by a thin but dense connective tissue layer, which connects the cremasteric fibres in a fascia—the fascia cremasterica.

The lower muscular fibres of the internal oblique arise from the deep surface of Poupart's ligament, as low down, in a well-formed subject, as the lower third of Poupart's ligament, covering over the transversalis fibres, and also the deep abdominal ring. In a muscular man they are frequently arranged in a thicker layer at this point, so as, with the subjacent

FIG. 2.

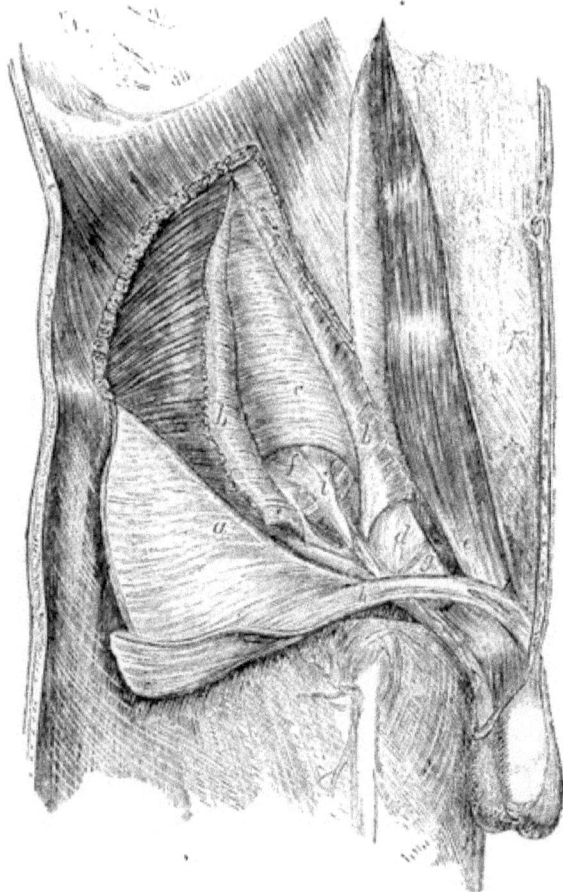

Parts of Inguinal Hernia : deep dissection, conjoined tendon (*d*, *b*), cremaster (*i*), and infundibar fascia (*f*) shown.

cremasteric fibres, to form an elevation or swelling, visible externally, so as to resemble a bubonocele somewhat. This swelling may be observed markedly emphasised in the well-known athletic statue, the Farnese Hercules. The lower fibres of the muscle arch closely over the cord at the deep ring—blended, more or less, with those of transversalis muscle which accompany them, reaching nearly as far down, and separated only by the small intercostal vessels and nerves, with the plexus of the ilio-inguinal and ilio-hypogastric nerves. Inside the canal they unite in a dense tough fascia to form a conjoined tendon. The outer curved border of this tendon is continuous, and blended with the fascia which covers both the surfaces of the transversalis muscle, especially with the deeper layer called the fascia transversalis. The muscular fibres of the internal oblique are inserted in strong bundles upon its superficial surface, so as to cover it in muscular males.

If the forefinger be carried along the canal in a hernial subject, and passed as far inwards as possible, so as to lift the muscular layers, the conjoined tendon can be raised upon it, along with the combined oblique and transversalis muscles, presenting a salient crescentic margin, easily recognised by the finger.

The cremasteric muscular fibres arise externally from the lower third, or rather more, of the deep surface of Poupart's ligament ; the higher fibres passing up behind the internal oblique, and connected with and sometimes receiving fibres from the transversalis muscle.

Below and internally the fibres spread out in a fan-like way, the upper ones arching over the cord to be implanted, continuous with those of the internal oblique, into the surface of the conjoined tendon. The lower two-thirds of the fibres pass out through

the superficial ring upon the cord, forming a second
sleeve-like investment, in combination with the connecting tissue.   The inner ones form a more or less
perfect series of loops, the recurrent fibres becoming
lost in the fascia about the angle of the pubis ;
while the outer or lowest are lost upon the fascia
investing the tunica vaginalis.   When the super-
jacent fascia intercolumnaris is opened, the finger can
easily separate the cremaster fascia from its connections, and invaginate it into the canal as high as the
deep ring.

At the outer border of the rectus abdominalis
muscle the conjoined tendon splits, a little above the
pubis, to form the sheath of that muscle.   The
separation of the layers is marked externally by a
crescentic depression; "the linea semilunaris," reaching from the pubic spine as high as the junction of
the eighth and ninth rib cartilages, into which it is
implanted.   Its posterior layer covers the deep
surface of the rectus muscle as far as the linea alba,
except at the lower fourth or fifth, where the muscle
is in contact with the fascia transversalis.   A curved
margin is here seen, called the fold of Douglas, under
which the epigastric vessels pass into the rectus muscle.

In a perfectly formed subject the outer border
of the rectus muscle forms a deep convex curve
outwards from the crest of the pubis, so as to
bring it into pretty close relation to the deep
ring, and to form a protection against hernial protrusion in this place.   The deep abdominal ring is
an oval opening in the fascia transversalis, placed
about three-fourths of an inch above the centre of
Poupart's ligament, with its longer diameter directed
upwards and inwards.

The opening is covered in front by a sleeve-like prolongation of the fascia transversalis, the borders of which give off over the cord a close investment of fascia, the fascia spermatica interna, or infundibuliform

FIG. 3.

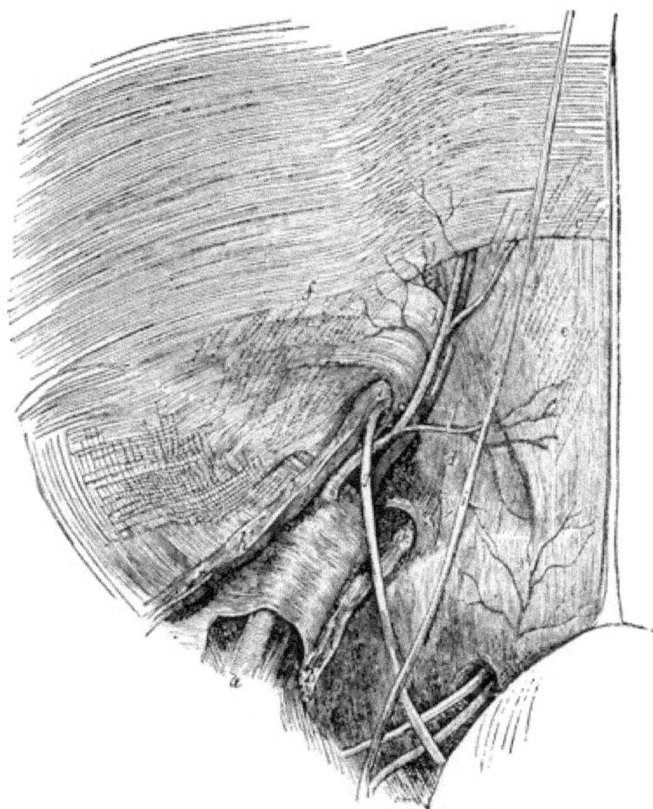

Parts of Hernia, dissected from within : Peritoneum removed, spermatic duct (g), epigastric vessels, and hypogastric cord (i) shown.

fascia ; the latter name from its funnel-shaped appearance.   Like the upper opening of a coat-sleeve, the aperture of the deep ring is seen only from its internal or deep aspect, presenting, when the peritoneum is stripped off, a sharp crescentic margin

C

internally, over which the spermatic duct can be seen to curve acutely. At this point, the spermatic vessels, lymphatics, and vaso-motor nerves first join the duct, lying on its outer and upper side, and bound to it by a connective tissue, the remains of the funicular prolongation of the peritoneum, which forms in early fœtal life the continuation of the tube of the tunica vaginalis. It is lost below on the outer surface of the tunica vaginalis when this is completely shut off from the peritoneum, and helps to form the spermatic cord.

When this has occurred, after the complete descent of the testicle, about the eighth month of intra-uterine life, the connective tissue becomes strengthened and connected with the subperitoneal fascia—a cicatricial dimple on the peritoneal surface, forming a strong resisting barrier against the first protrusion of a rupture into the inguinal canal. On this aspect of the hinder wall of the canal is seen, in the subperitoneal fascia, a tolerably sized artery, given off from the external iliac just above Poupart's ligament, the deep epigastric artery. It passes upwards and inwards, to enter the sheath of the rectus at the outer part of the fold of Douglas, accompanied by two veins, one on each side, which join before they open into the external iliac vein. Between these vessels and the outer edge of the rectus is formed a triangular space, with its base towards the inner third of Poupart's ligament (the triangle of Hesselbach). This is covered by the fascia transversalis, closely adherent to the deep surface of the conjoined tendon of the internal oblique and transversalis muscles. In front of the latter, at the lower and inner angle of the aforesaid triangle of Hesselbach, can be seen a triangular arrangement of tendinous fibres, which can be traced from the internal pillar of the superficial ring of the opposite side of the body, across the median line, or linea alba, downwards and outwards to the inner fourth of Poupart's ligament, with which, and with the conjoined tendon, it is intimately blended at the pectineal line.

Across the triangle of Hesselbach, on the deep or peritoneal surface, the obliterated hypogastric artery, forming part of the superior false ligament of the bladder, is seen to cross upwards and inwards towards the umbilicus. On each side of it is a loose pouch of peritoneum, which in direct inguinal hernia forms a sac for the rupture, and also affords material for the enlargement of the sac in the oblique variety.

The inguinal canal is a passage from the deep to the superficial ring, one inch and a half to two inches in length, and directed obliquely in a double sense,—viz. oblique from above downwards and inwards, and oblique also from the deep to the superficial opening. It is closed in a valvular way by the close apposition and areolar connection of the deep and superficial walls or boundaries. This adhesion by connective tissue is torn or stretched in the formation of a hernia, or prevented altogether by the persistence of the funicular process of peritoneum in infancy, and it is one of the most important objects of a radical cure of hernia to restore this adhesion and reinstate the valvular action of the walls of the canal if possible.

The deep wall is formed by the apposition and union of the layers of aponeurotic fascia—namely, the fascia transversalis, blended with the conjoined tendon of the internal oblique and transversalis muscles, and at the lower third with the fascia triangularis, passing from the external oblique of the opposite side to the lower third of Poupart's ligament, and covering the lower tendinous origins of the rectus abdominis muscle.

The anterior wall is composed mainly of the aponeurosis of the external oblique muscle, where it forms the outer pillar of the superficial ring. At the outer two-thirds, beneath this, are the muscular fibres of origin of the internal oblique muscle, arching over the canal to unite with the edge of the conjoined tendon. Closely skirting the deep ring beneath this are the lower fibres of the transversalis muscle, connected with it by a tough fascia. At the inner or lower half are the fibres of the cremaster muscle, also closely attached by its connecting fascia to the conjoined tendon and fascia triangularis.

Below the canal is the grooved upper edge of Poupart's ligament, and above it are the arching fibres of the internal oblique and transversalis muscles. When pressure from within is made by the viscera the hinder wall is pressed forwards against the front wall, and the passage is thus fortified against dilatation. At the same time the elastic peritoneum and its fascia, closing strongly the deep ring, prevents the omentum or bowel from forming a depression, which would give it a hold and purchase wherewith to dilate the opening and canal.

Formation of Oblique Inguinal Hernia.— If this closed ring yield from imperfect formation or want of tensile power, a depression is formed outside the epigastric vessels. This affords a prominent and resisting line or border, so that a hernial dimple or shallow sac is formed, permitting the viscera to protrude and press against the internal and external oblique fibres, and to bulge forward the anterior wall of the canal. Thus a hernial protrusion is .commenced (bubonocele), which speedily, under muscular effort, turns downwards along the canal, rips up and

FIG. 4.

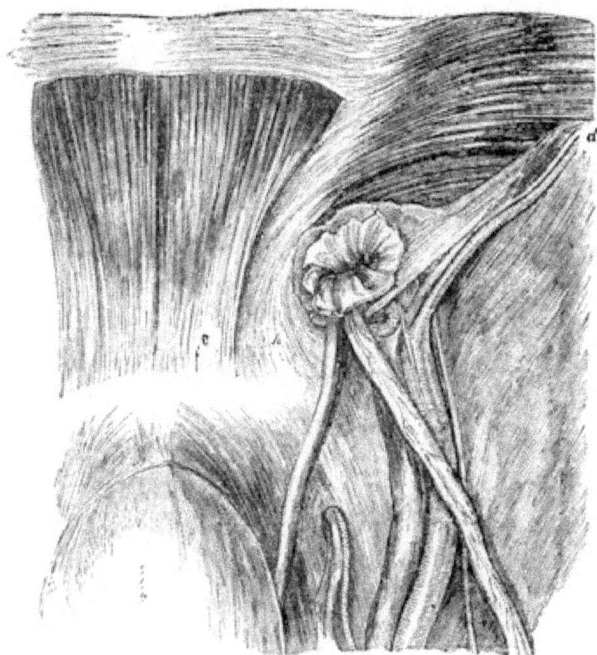

Oblique Inguinal Hernia: dissection from within; sac cut off from peritoneum : (*h*) conjoined tendon, and (*c*) rectus tendon.

stretches the connective tissue binding the walls together.

When, by a late and imperfect development of the gubernaculum or other causes, the testis has not descended in the eighth month of fœtal life, a sac is already formed by the funicular process of peritoneum.

In these cases the internal oblique and transversalis muscles, as well as the cremasteric fibres, are weak and imperfectly formed ; and the former does not extend at its origin so far down as it ought, sometimes occupying only the outer third of Poupart's ligament.   Thus but a feeble resistance is afforded to the descent of a hernia at the moment of the greatest tension. At the same time, the superficial ring may be larger than usual, in consequence of being occupied by the testis in its slow and delayed descent.   The intercolumnar fibres are feeble and yielding, and afford but little resistance to the separation of the pillars of the ring.   The external aspect of the patient's groin in these cases is weak and bulgy, and the deficiency of development of the muscular fibres sufficiently evident.   The rectus muscle is narrow, and does not arch boldly outwards to close up and protect the groin-area.

The testis may be in the deep ring or the canal, in the superficial ring, or just upon the pubic angle. In such cases a hernia almost invariably follows the descent of the testis.   When the gland is adherent in the canal, the epididymis may be drawn out and stretched, so as to be altogether below the testicle, with a pouch of peritoneum (the aborted tunica vaginalis) quite below both, and simulating the testis externally to the eye of the surgeon.   The contour of the external genital organs in these patients is usually rounded and infantile, sometimes stunted.

The persistence of a small funicular process of peritoneum in the canal may result, even later than the infantile period of life, in a hernia ; first the omentum, and then the bowel, dilating the canal down to its imperfect closure at some point above the testicle, where it forms a large tunica vaginalis. In children such cases may thus have the fundus of the hernial sac invaginated into the large tunica vaginalis, giving a double serous covering or sac to a hernia. This is described in text-books as an infantile hernia, and is characterised as presenting three layers of serous membrane for division before the contents of the sac are reached. It must be considered as simply a variety of congenital hernia. In the ordinary congenital hernia, the funicular offset of the peritoneum is not closed up at all, but the hernial contents descend and lie directly upon, and may reach beyond, the testicle. Such a hernia is characterised by its rapid progress downwards into the scrotum, having only to dilate the funicular tube of serous membrane, and to overcome but little resistance from the already open inguinal canal.

It is to a late descent of the testis, to an imperfect closure or a mere narrowing of the funicular process, and to the rapid development of the small intestine and omentum, that I attribute the occurrence of inguinal hernia in infants and young children. In adult cases also there is frequently a history of some sort of a swelling about the scrotum in childhood.

We have seen that, out of 370 cases of inguinal hernia, as many as 123 could be traced to a congenital origin. In a great proportion of these, ruptures had been present in the males of the family —in the fathers, uncles, or grandfathers; showing the hereditary nature of the cause. In many instances there were open rings and a bulgy loose formation of the groin where no actual rupture was present, revealing the imperfect development of the muscular apparatus and surroundings of the spermatic cord. In many others also there was an imperfect development of the testicle on one or both sides, most frequently on the right side. The majority of cases of inguinal hernia being also on that aside, tends to prove that the cause of this variety of hernia here assigned is the true one.

When the abnormally feeble barrier at the deep ring has once begun to yield, other agents tend to promote the formation of a complete hernia. The bubonocele pushes forward the anterior wall opposite the deep ring, weakened by the want of development of the lower part of the internal oblique and transversalis muscles. The continued action of the latter muscle, forming a horizontal muscular constriction on the abdomen, forces the intestines down upon the weakened ring. At the same time the contraction of the recti muscles keeps them in the same position, and by the backward pressure upon the sheath forces the deep wall of the inguinal canal backward, and tends to open up the valve of the canal, while the sac finally slips down and emerges through the imperfectly closed superficial ring, and then opens up the channel of the spermatic cord down into the scrotum, distending the dartoid pouch-like investment.

In a condition of oblique inguinal hernia, the cord lies first below and to the inner side of the neck of the sack, the fundus of which is lodged, in the earlier stages of bubonocele, in a deep groove behind the outer pillar of the superficial ring, and above Poupart's ligament. As it proceeds along the canal the cord crosses obliquely behind the sac of the hernia, and finally lies behind and to the outside of the complete hernia, in the groove on the outer pillar of the ring, external to the pubic spine. The

sac, when distended, fills up the canal, adapting itself
to the shape of the passage, so that there are usually
found three slight constrictions: namely, one at the
neck of the sac, one at the lower margin of the

FIG. 5.

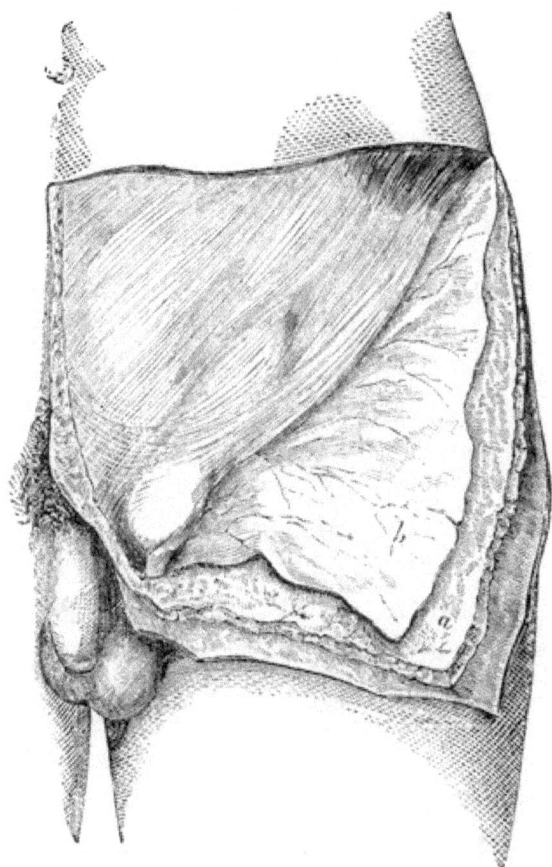

Superficial Dissection of Bubonocele.

internal oblique, and one at the superficial ring.
The first is usually the most marked, and is often
attended with a thickening of the neck of the sac
itself, which may constitute the strangulating portion
of a hernia. The constriction may be at any of

the three points indicated, but is most common at the first.

In the formation of a direct inguinal hernia, much less common, the deep ring has no share. It resists the applied force, while the conjoined tendon and the other layers of the hinder wall of the canal below the epigastric vessels, and between them and the hypogastric cord, yield before the loose peritoneal protrusion, or are, less commonly, split and really ruptured by the muscular force applied above. Pushed before the intestines, the loose and yielding peritoneal sac passes between those vessels, or internal to both of them. The spermatic cord usually lies to the outer side of the sac of a direct hernia, all the way along the canal.

Not unfrequently it was found, when operating to remove the sac, that the constituents of the cord were pushed in front of the sac, and sometimes spread out and separated from each other, and were even projecting into the cavity of the sac.

This constitutes a dangerous and perplexing condition to the operator when removing the sac of a large hernia. The distinction usually drawn between the direct or internal and oblique or external hernia from the relative position of the epigastric artery to the neck of the sac, is difficult or impossible to detect in the operation for the radical cure; and even in that for strangulation it rarely affords any guide to the operator, to lead him to depart from the rule of cutting upwards. An oblique hernia of old standing may present the appearance of a direct one from the rings being drawn nearer to one another by the enlargement of the deep ring, and the neck of the sac may be in close contact with the edge of the rectus muscle.   No trouble has ever been caused in any of my operations by the epigastric vessels, and no hæmorrhage has ever followed the operation.

The structures which the surgeon must divide to reach the sac of the oblique variety, outside the superficial ring, are as follows, namely : the integuments, including skin and two layers of superficial fascia, with the small cutaneous vessels and nerves ; the fascia intercolumnaris ; the cremasteric fascia, recognised by the presence of muscular loops, with the cremasteric nerve-vessels ; the fascia infundibuliformis *vel* fascia propria ; and then, covered by a little adipose connective tissue (fascia subperitonalis), the sac of the hernia.

The structures divided in cutting down to a direct hernia proper are the same, with the substitution of the fascia transversalis for the infundibuliform, and the conjoined tendon for the cremasteric fascia. The conjoined tendon is sometimes split by the hernial sac, and does not cover it. A few cremasteric fibres may, however, be seen in muscular subjects lying over the direct hernial sac, when this is placed close internal to the epigastric artery.

The occurrence of a double sac, one through the deep ring and another through the conjoined tendon, is rare, but must be kept in mind.

A direct hernia is, more commonly, the consequence of an overpowering muscular force, rapid in its action and effects, and resulting in a distinct tear or rupture of fibres, than the other form of inguinal hernia, which is the result of gradual yielding.   It is most frequent in adults whose muscular power is disproportionate to the existing strength of the fibrous structures.   The neck of the sac lies close to the edge of the rectus, and there is no oblique valve-action capable of being restored by operation. Sometimes the hernial sac emerges from the canal by splitting the fibres of the outer or inner pillar of the superficial ring, and may thus, in the former case, be placed in close proximity to the femoral vessels.

FIG. 6.

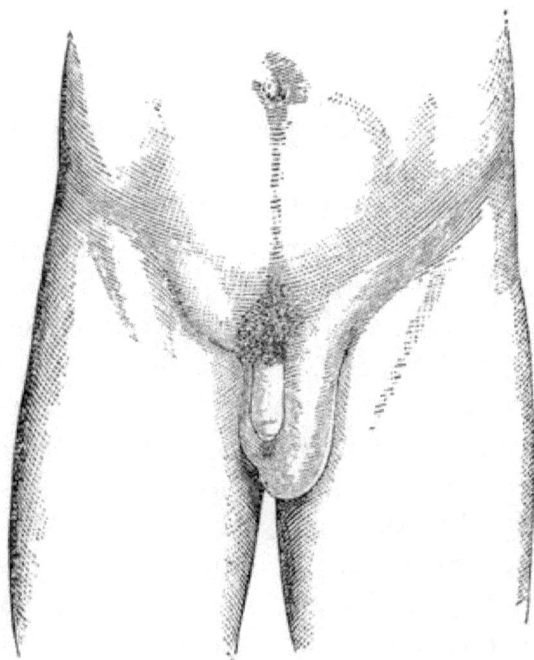

Right,—direct Bubonocele, and left,—oblique scrotal Hernia.

Industrial and gymnastic exercises favourable to the prevention and cure of hernia are those which tend to strengthen the abdominal muscles and fascia, such as hanging and swinging by the hands upon the trapeze, swaying from one hand to the other, swarming up a pole or up a rope or rope-

ladder by the hands, or any motion or posture which
stretches out the body and increases the tension of
Poupart's ligament.

But force applied with the legs bent on the body
and the groin structures relaxed is injurious, and
productive of rupture; as in rowing, tricycling, lifting
heavy weights, and straining with the diaphragm.
In all these, Poupart's ligament and the structures
connected with it are relaxed ; while the transversalis muscles exercise a constricting power upon the
abdominal viscera, forcing them downwards upon
the relaxed groin structures, which give way before
the force, and produce rupture. The rectus also,
when contracting, presses backwards its sheath and
conjoined tendon, and so tends to open up the canal.

The contents of an inguinal hernial sac vary very
much, and sometimes perplex the surgeon both in
diagnosis and operation. The funicular process
may be patulous throughout, though too narrow to
admit of either bowel or omentum. A serious
effusion may be present, which forms a translucent
tumour like hydrocele. This may disappear on the
patient lying down, and may be reduced by gentle
pressure more or less slowly, according to the size
of the tube communicating with the peritoneum.
This has been called a diffused hydrocele of the
tunica vaginalis. It is usually present in children.
Before it become absorbed, the canal of the funicular process may become closed at some point in
the canal ; and a doughy, loosely fluctuating scrotal
tumour may be present, which does not disappear
on pressure or lying down. This is sometimes
called a windy rupture : neither bowel nor omentum
can be detected.

When the canal is closed up imperfectly in two places or more, a tumour may be formed on the cord, elastic and irreducible, from fluid effusion into the canal between the obstructed points, and the complaint then constitutes one form of encysted hydrocele of the cord. At the same time a pouch may exist at the deep ring, forming a true hernial protrusion.

Occasionally I have found in a hernial sac one or more of the bodies known as peritoneal concretions. Round as a billiard-ball, and sometimes of considerable size, such concretions seem to be formed by a deposit of albumen in layers around a detached appendix epiploica, and may slip up and down from the sac to the abdomen.

The hernia may be reducible from enlargement by fatty fibroid deposits in the omentum, constituting one form of incarcerated omentum.

Or the omentum may be adherent to the bowel, or one or both of these to the sac. If such adhesions be recent and not extensive, they may be detached in an operation for cure, the omentum removed, and the bowel returned, all bleeding points being carefully ligatured. Sometimes I have found a secondary smaller sac budding from the primary hernial one, and filled with a hard and convoluted mass of adherent omentum. In some cases these have been found which in size and shape feel like a testicle, the real gland being obscured by the mass. They have not, of course, the testicular feel to the patient on pressure, and this forms something of a guide when the patient can be trusted to help in the diagnosis. Lastly, there may be present in an old and large rupture, on either side, a portion of large intestine, the *caput cæcum coli*, a part of the bladder, uterus, and ovaries. I have never found either stomach, spleen, or kidneys in a hernial sac of any kind, but such cases are recorded.

Adhesions and constrictions of the sac or contents of a hernia may give rise to strangulation, which illustrates the propriety of always, as a rule of operation on strangulated hernia, opening and examining the interior of the sac. This is still more forcibly inculcated when ulceration sphacelus or fæcal extravasation might be present.

The sac of the hernia may be so thin and delicate that the necessity for removing it is not apparent. Twisting and stitching up the neck along its whole length, together with the parietes and openings of the canal, have in 261 cases out of 305 been found sufficient to cure cases of reducible hernia, operated on when none but reducible cases were judged proper for the radical cure. In the other cases the sac was removed.

LECTURE II.

Operation by the Subcutaneous Method.—Causes of Failure in the Operation.—Summary of Cases of Inguinal Hernia operated on for Radical Cure.—Successes and Failures of the operation.—Table of ascertained Duration of Cure.—Number of ascertained Failures.— Hernia Complicated with Undescended Testis.—Transplantation of Testis into Scrotum.—Formation of " Tunica Vaginalis."

IN the subcutaneous method of operating, an incision is made in the upper part of the scrotum, where the tegumentary structures are so movable and elastic that the incision can, if necessary, be drawn upwards so as to expose the superficial abdominal ring ; and can be enlarged, if it be desirable to remove the sac, to an extent sufficient for this purpose.' The length of this incision rarely exceeds, when tension is removed and the dartos has resumed its normal degree of contraction, one and a half inches. It will now be easy to detect with the finger the pillars of the superficial ring, the lower border of the internal oblique and transversalis muscles, the borders of the deep ring, the spermatic cord, and the outer edge of the rectus abdominis muscle. The finger can be passed into the deep ring, so as to lift forward the internal oblique and transversalis muscles, and feel internally the raised edge of the conjoined tendon.

The first material I employed in stitching up the canal was hempen thread, fastened over a wooden compress. This was used in seventeen cases. In the small cases of children, a pair of rectangular pins, similar to hare-lip pins at the point, but bent at a right angle near the other end, with a loop like that of a safety-pin, to lock into its fellow, were used (in forty-eight cases). Then stout copper-wire, silvered, was employed, with a view to prevent suppuration, and to afford a straight direction to the tract, so as to drain the part effectually (in two hundred and forty cases). Latterly I have used a stout piece of kangaroo- deer- or ox-tendon, well antisepticized in carbolized oil, and softened just before using by soaking in 1 to 40 carbolic lotion. The advantage of this is, that there is no necessity for disturbing the wound by removing the buried suture, as in the case of the wire and other methods.

All these modifications are essentially alike as regards the structures which were traversed and in-

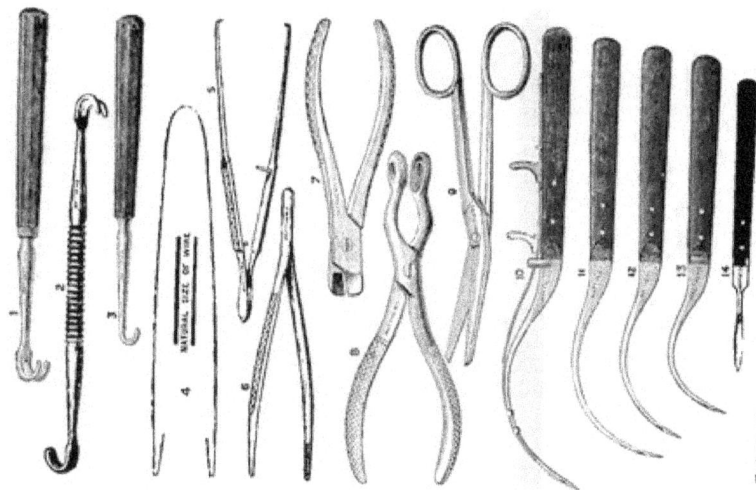

INSTRUMENTS USED IN MR. WOODS OPERATIONS
FOR HERNIA.

NATURAL SIZE OF WIRE

To face page 31.

cluded in the suture. The variations in manipulation were merely those rendered necessary by the nature of the material employed for the ligature. When wire was employed, it was usually removed after a week or ten days. Hardly any discharge followed its use and withdrawal. The induration of the parts around and within the inguinal canal was usually marked, but its duration was not so long as in the case of the buried tendon suture. It was soon found that this induration, which was formerly considered by surgeons as important, did not remain as a permanent barrier against a return of the hernia. Such a barrier can only be obtained by a more intimate and extensive adhesion between the sides of the canal.

Operation by the Subcutaneous Method.—

FIG. 7.

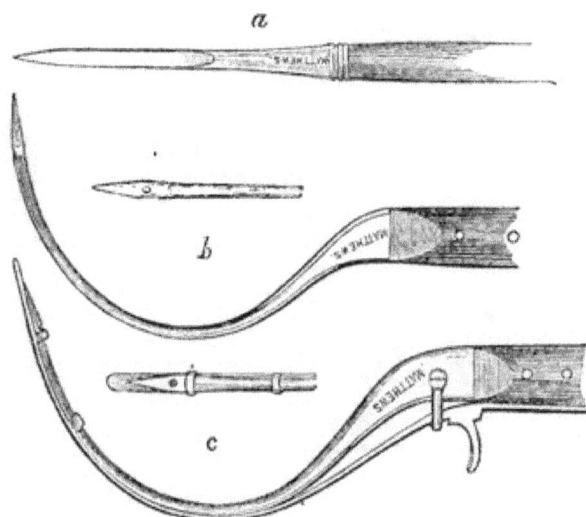

Knife (*a*) and needles (*b*, *c*) used in Wood's operation.

A tenotomy knife, a semicircular needle, mounted on a stout handle, flattened at the eye, with a sharp

D

point and blunt shoulders, formed so as to slip along
the front of the curved forefinger, with about a foot
of tendon or wire as thick as stout twine, are the instruments necessary for the operation. If the sac is
to be removed, a pair of blunt-pointed bent scissors
will be useful, and a double hook or two to hold the
edges of the wound apart.

FIG. 8.

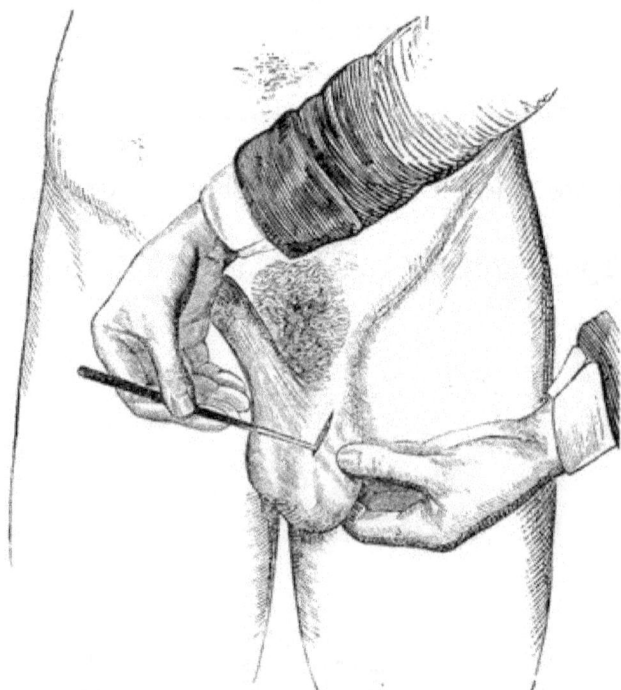

First step in operation for the Radical Cure of Inguinal Hernia

The parts being carefully shaved and washed,
together with the instruments and hands of the
operator and assistant, with a 1 to 20 solution
of carbolic acid, the hernia returned, the patient
anæsthetised, and the spray (if considered important)
in action, an incision with the tenotomy knife, about

three-fourths of an inch long and oblique in direction, is made over the cord just below the pubic crest, through the tegumentary coverings down to the sac. A small artery—the external pubic—may require section and ligature at both ends with small catgut. The forefinger is then passed into the canal, carrying

FIG. 9.

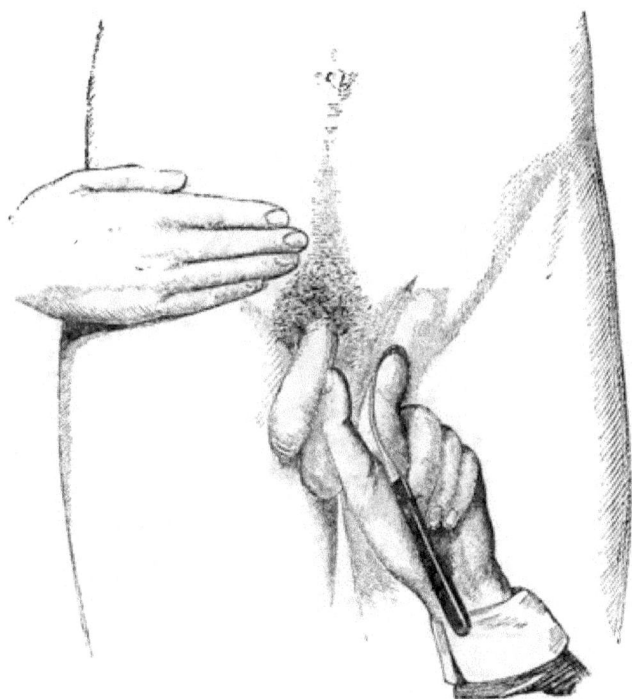

Showing the sliding of Skin and Fascia inwards after transfixing conjoined Tendon.

the sac invaginated before it, up to the deep ring, behind the internal oblique and transversalis muscles, which can be felt and seen to be lifted up by the finger. At the inner border the edge of the deep ring and conjoined tendon can be felt, and the finger passed behind them. Along the finger is then carried

D 2

the needle, until its point can be felt behind these
structures, through which it is then pushed.   When
its point raises the skin, the latter is drawn well
over to the inner side before the needle is pushed
through it.   The tendon is then passed through the
eye of the needle (or, if wire be used, a bend at the
end of it is hooked on), and withdrawn with the

FIG. 10.

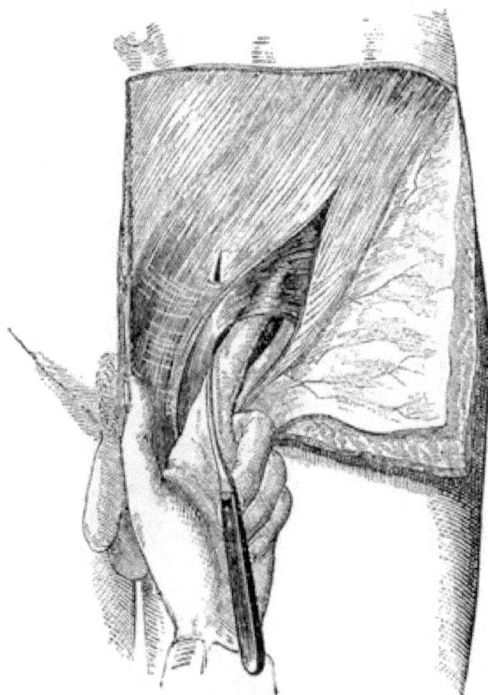

Showing application of needle to conjoined Tendon.

needle through the scrotal puncture, and then detached.   The finger is then passed behind Poupart's
ligament, the spermatic cord felt for in its groove,
and pushed aside.   The point of the finger is then
carried close to the ligament, along the groove of the
outer pillar of the superficial ring till it is opposite to

the deep ring, the needle passed along it as before
and made to pierce the aponeurotic fibres. The skin
is then drawn outwards, so that the needle can be
passed through the same puncture as before. The
other end of the ligature is then secured to it, drawn

FIG. 11.

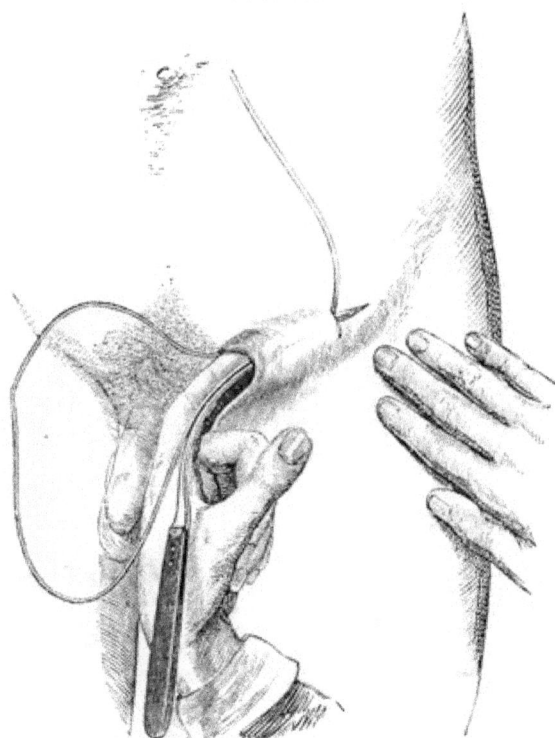

Showing the sliding of Fascia outwards after piercing
Poupart's Ligament.

again through the scrotal puncture, and detached.
The needle is then carried across behind the sac,
between it and the cord. The latter can be isolated,
as in tying the veins in varicocele, without difficulty.
The inner end of the ligature is then connected with
the eye of the needle, and drawn through. In a

large case, especially if the rupture be a direct one,
the needle is, lastly, to be passed through the end of
Poupart's ligament, just above the pubic spine, and
then carried through the inner pillar of the ring and
triangular fascia, close to the os pubis, at the edge of

FIG. 12.

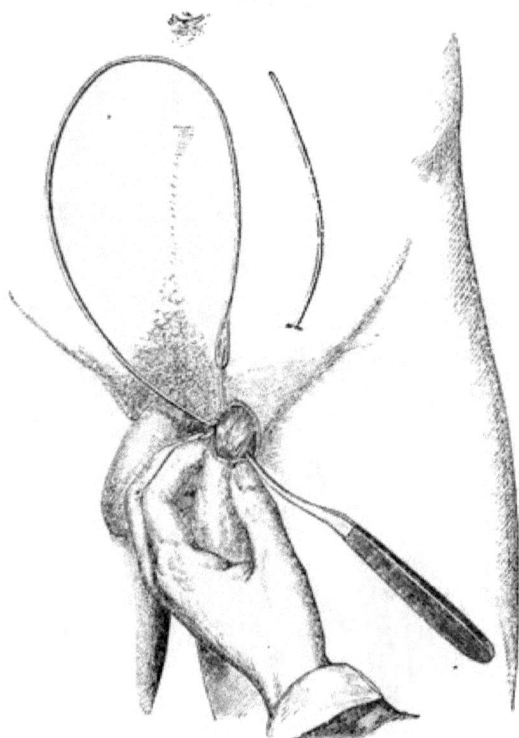

Transfixion and taking up the sac of the Hernia at
the Scrotal Puncture.

the rectus muscle. The outer end of the ligature is
then connected, and drawn across so as to lace up
the canal like a boot.

In the case of tendon-ligature being used, it in
now to be braced up tightly, tied in a well-secured
surgeon's knot, cut off close, and buried in the wound.

If wire be used, a loop is left at the upper groin-puncture; the lower ends are twisted down into the scrotal aperture, the upper loop drawn upon so as to tighten the wire, and held firm by two or three twists down into the puncture. The bight of the upper loop is then bent downwards to meet the ends bent

FIG. 13.

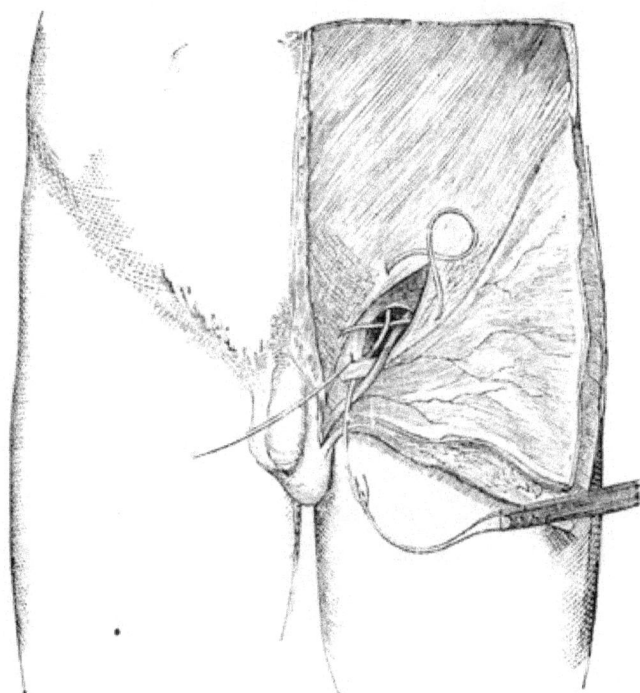

Dissection of Groin : showing the application of the wire or tendon lacing up the Inguinal Canal.

upwards; these are curved into the form of a hook to fasten on to the loop. A firm pad of lint is placed on the skin under the arch thus formed; and a broad spica bandage secures the whole, and exercises suffi-cient pressure to keep the wire firmly in contact with the hinder wall of the canal.

In wire cases, no drainage tube or antiseptic dressing is required.   The wire acts as a straight and

FIG. 14.

perfect down drain : the operation is subcutaneous, and the
wound rarely even suppurates.

In the case of tendon being
used, a drainage tube should be
placed, reaching from the superficial ring through the scrotal
puncture, and the gauze dressing
applied in the usual way by a
double spica bandage, with a piece
of jacquinette, through which
the penis is passed, placed over
all to keep off urine from the absorbent dressing.

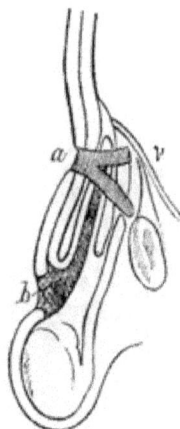

Showing lines of union
in track of wire.

If the scrotal opening be made
larger, in order to remove the
sac, some stitches should be placed pretty
close together above the drainage tube.   In such a
case the scrotal opening can be drawn up and
stretched, so as to allow the needle to pass out and
in through it, instead of through a separate groin
puncture.   Thus the whole operation may be done
through a scrotal opening of the length of $1\frac{1}{2}$ to 2
inches.   The patient should be placed in bed in a
half-sitting posture, in a bed-chair, with the knees
drawn up over a bolster.   Sometimes it is necessary
to use the catheter a few times after the operation.
Usually there is no need of this, and the patient is free
from pain after the first twelve hours.   To prevent
pain altogether, a morphia suppository should be
placed in the rectum at the time of the operation.

In a week or ten days, according to the amount of

action, the wire may be withdrawn. By this time both the ends of the bent wire lie in the same wound-track. The lower twist is first untwisted, and the ends stretched by extension to efface the spiral twist. When the ends of the wire are cut off close below, the wire usually comes out in one piece by traction upon the loop above. If there be any difficulty the loop may be divided, and the ends dealt with separately. The great advantage of using carbolized absorbent tendon is that this proceeding, somewhat painful, is avoided.

If this operation be properly done, it fulfils the following requirements for the permanent cure of inguinal hernia.

1. The deep ring and hernial opening are closed flush with the peritoneum, while the internal oblique and transversalis muscles, and the external oblique aponeurosis, are united to each other and to the deep hernial opening and mouth of the sac, so as to close, sustain, and support it against a fresh protrusion.

2. The conjoined tendon, forming the hinder or deep wall of the canal, is united to Poupart's ligament, close upon and over the spermatic cord and twisted sac. Thus the valve action of the canal walls is restored, and the deep ring supported from below. The muscular and aponeurotic layers between which the canal lies are bound together by adhesion, where they had been separated by the hernia.

3. The pillars of the superficial ring are laced up like a boot, supplementing the weakened arciform fascia, supporting the other adhesions, and forming a third line of defence against a renewal of the protrusion. There is no permanent invagination of

fascia after the purposes of the operation itself are
fulfilled. The firm mass of the material, which afterwards becomes apparent, is composed of fibrinous
effusion, which contracts and hardens like any other
cicatrix ; and this blending of the three layers of
suture forms the barrier upon which the surgeon
must rely.

Causes of Failure in the Operation.—The
most common is failure to secure the sides of the
deep ring by not planting the suture close to its

FIG. 15.

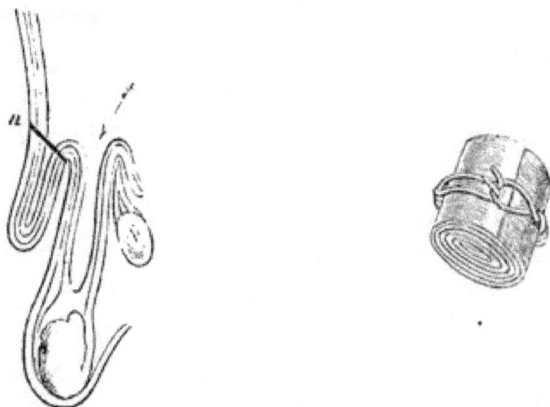

Showing, to the left, the Invagination for the application
of Ligature at the Deep Ring ; and, to the right, the
wire folded over pad.

edges. Here fear of damage to the deep epigastric
artery acts sometimes to the disadvantage of the case.
These vessels, however, are so loosely attached to the
tissues, and so movable, as to be cut with difficulty in a
puncture with a blunt shouldered needle. They slip
out of the way, and though doubtless they have been.
often included in the grasp of the ligature, they have
never given rise to any trouble or secondary

hæmorrhage in the whole 414 operations. Next,
the operator may fail to secure the conjoined tendon
properly, or it may give way prematurely to the action
of the ligature. Then, from fear of wounding the
spermatic chord, the ligature may not embrace the
tendon sufficiently closely, and may not be passed
deeply enough through Poupart's ligament. The
hernial groove here formed may not be included in
its grasp, so the hernia may again creep along the
unclosed canal (see Fig. 15). Then the pillars of the
superficial ring may not be closely and continuously
united along their whole length to each other and to
the triangular fascia. Lastly, adhesions formed when
the patient is in a feeble state of health, or in a permanently lax and weak condition of the aponeurotic
structures, may yield and give way, as they may do
in other operations for prolapse of any kind, by repeated and constant stretching. In such cases a
preventive truss will be afterwards required.

Sources of Danger in the Operation.—In
the first place is the possibility of a puncture or
wound of the bowel in placing the deep sutures.
By keeping the needle upon the front of the curved
finger, placed in the proper position, this can be
easily avoided. In point of fact, this accident has
not once happened in my own hands. Next, it is
possible that the femoral or iliac vessels may be
damaged. By placing the finger in front of the
vessels, and lifting up Poupart's ligament well from
them, I have avoided this accident. It has never
occurred in my hands ; nor, as before mentioned,
has any trouble arisen from the epigastric vessels.
The cord may be damaged, and the spermatic duct
included in the ligature or obstructed by its pressure.

In two cases out of 414 the testicle has become atrophied ; in one, in consequence of the pressure of a steel clamp to the ends of the wire, and in another from an abscess forming in the 'gland. In two others, being found atrophied, it was intentionally removed in the operation.

Summary of Cases of Inguinal Hernia operated on for the Radical Cure.—I will first take the cases of reducible hernia, which for some years were the only class of cases upon which I deemed it advisable to operate for the radical cure.

In my earlier attempts, stout hempen ligature thread was used, applied subcutaneously, and secured over a compress. Of these there were seventeen cases, with one death from pyæmia.

The pin operation was chiefly done on infants and children with large uncontrollable ruptures. Of these there were forty-eight cases, one a double operation, making forty-nine operations, with two deaths, from erysipelas and peritonitis ; the last was set up by the pressure of a strong truss just before the operation upon a knuckle of bowel, which was found after death to be the focus of inflammation on the opposite side of the abdomen. This was published in the *British Medical Journal* at the time.

The subcutaneous wire operation was performed in 252 cases, of which nine were double operations (on both sides), and eleven were second operations (the first not having succeeded). With these, the number of operations, as distinguished from cases, was 273.

Two hundred of these were done consecutively, without a single death or unpleasant symptom. In all these cases the sac was not removed, but transfixed in several places, twisted, and tied firmly. The cases were selected mainly from young male adults, or adolescents, in good health, which may partly account for the singular immunity from serious symptoms in two hundred consecutive cases. Five of them only were females, all young adults or children. Out of the whole 273 there were four deaths : one from tetanus, one from delirium tremens, and two from broncho-pneumonia.

The mortality in these three variations of the operation, taken together, was seven deaths out of 339 operations, or about 2 per cent., and half of them were from hospital causes, some of which have been of late years entirely abolished. Taking the wire operations alone, there were 273 with four deaths—about 1¾ per cent.—none of which could be strictly attributed to the special conditions or results of the operation.

The Successes and Failures of the Operation.—In an operation of this kind there are special difficulties in estimating and ascertaining the proportions of failures to successful cases. A certain proportion of the failures can be verified by the patients returning for further aid to the surgeon.

If the operation be imperfectly performed, the failure becomes manifest before the patient leaves the care of the surgeon, and can be noted. But in a considerable proportion of hospital cases the patient does not reappear after the first convalescence. My experience is, that he is more likely to come back, if the operation be unsuccessful, in search of further relief. However carefully the addresses of hospital patients may be kept, their wandering habits and frequent removals render it impossible to keep them in view, and to follow up the cases, for many years.

I have found that in the greater proportion of the failures the rupture returns before the end of the first year. When the operation is properly done, but the weakness of the abdominal structures inherent in a patient's tissues results in a slow yielding of the parts and the formation of a fresh sac, the rupture may not become apparent until some time during the second year.

It has often been the custom of surgeons, too anxious for the appearance of success, to set down and put forth as cures cases which have been examined only a few weeks or months after the operation. The short-sightedness of placing the operator's reputation upon such a foundation has often struck me, and to avoid it I have, in the list which my hearers have placed before them of the results of ninety-six successful cases (furnishing ninety-eight operations), after subcutaneous operation, carefully excluded all cases which have not been verified two years after the operation.

It will be seen in the last column of the table that the first case has been watched and examined in public during a period of not less than twenty-five years.

Two cases have been in view for twenty-three years (one of these cases is present for examination), one for nineteen, another for seventeen, three for sixteen, and so on, down to four years' duration. Of the last there are eleven cases; of three years' duration twenty-one cases, and of two years upwards of twenty-four cases. The dates of

TABLE OF ASCERTAINED DURATION OF CURE.

After Wood's Subcutaneous Operation for the Radical Cure of Reducible Inguinal Hernia.

No.	Initials of Name.	Age	Date of Operation.	Date when last seen or heard of.	Truss worn or not.	Duration of Cure. Abt. yrs	Remarks.
1	H. H.	22	Oct. 6, 1860	Jan. 5, 1885	No.	25	Shown frequently at King's Coll. Hosp. & Royal Med.-Chir. Soc.
2	J. B.	25	Jan. 11, 1862	Feb. 14, 1885	No.	23	Shown frequently at King's Coll. Hosp. & Royal Med.-Chir. Soc.
3	H. C.	28	Feb. 28, 1862	Jan. 1885	Light tr.	23	Indian service.
4	R. S.	6	Mar. 24, 1864	Mar. 5, 1885	No.	21	
5	H. W.	30	Nov. 1864	Jan. 1885	Light tr.	19	Fell down stairs in Nov. 1877, [side after 17 years without truss.
6	J. M.	25	June 25, 1860	Nov. 20, 1877	Light tr.	17	and ruptured himself again same
7	H. H.	16	June 29, 1859	July 16, 1875	Occasion.	16	Wears truss at hard work; engine-fitting,
8	C. C. T.	19	June 5, 1865	July 19, 1881	No.	16	
9	H. W.	26	June 21, 1865	April 30, 1881	Occasion.	16	Seen again quite recently.
10	F. H.	25	Nov. 8, 1871	Mar. 5, 1885	No.	14	Operated on three times; pilot.
11	M. B.	5	Oct. 1864	Dec. 1877	No.	13	
12	G. T.	15	Oct. 5, 1861	Dec. 9, 1874	No.	13	
13	W. A.	5	Oct. 6, 1864	Oct. 9, 1877	No.	13	
14	A. H. B.	50	Oct. 20, 1867	May 16, 1880	Light tr.	13	Heard from by letter.
15	C. T.	18	June 28, 1862	April 1874	No.	12	Seen many times.
16	M. S.	8	Nov. 14, 1872	Oct. 1884	No.	12	
17	J. B.	17	June 6, 1859	Oct. 5, 1870	No.	11½	
18	Capt. G.	24	April 6, 1858	Dec. 6, 1883	No.	11	Been in India eight years, riding hard.
19	J. C.	25	Sept. 5, 1863	Oct. 20, 1868	No.	10½	Died of phthisis; no return up to death.
20	T. L.	18	Sept. 5, 1862	May 6, 1873	...	10	
21	D. W.	34	July 5, 1876	June 10, 1872	No.	9	
22	A. P.	22	May 31, 1862	Jan. 15, 1885	No.	9	India; wears truss when riding.
23	W. R.	19	Oct. 19, 1872	Jan. 1870	No.	8	
24	G. H. F. J.	28	Feb. 13, 1877	Mar. 5, 1880	Light tr.	8	
25	L. E.	26	Mar. 30, 1878	Oct. 1884	No.	7	Double; both sides operated on.
26	J. P.	19	Mar. 1866	Feb. 10, 1885	No.	7	Double; both sides operated on.
27	Lieut. D.	26		Jan. 1873	No.	7	Heard by letter.
28	T. N.	7	July 4, 1863	Dec. 28, 1869	No.	6½	
29	G. N.	9	Oct. 3, 1863	Dec. 28, 1869	No.	6	
30	H. C.	20	Oct. 5, 1866	Nov. 1872	No.	6	Died of consumption; no return up to death.

TABLE OF ASCERTAINED DURATION OF CURE—*Continued.*

No.	Initials of Name.	Age	Date of Operation.	Date when last seen or heard of.	Truss worn or not.	Duration of Cure.	REMARKS.
31	G. R. A.	26	June 19, 1868	Sept. 1874	No.	6	Died of gastric fever; no return.
32	A. H.	22	Feb. 9, 1878	Jan. 25, 1884	No.	6	
33	J. P. N.	24	Dec. 13, 1870	April 5, 1876	Light tr.	5½	
34	G. V.	21	June 1860	Oct. 1865	No.	5	
35	W. B.	32	Mar. 12, 1862	May 1, 1867	Light tr.	5	
36	W. A.	24	Oct. 3, 1862	Nov. 1, 1867	...	5	
37	G. R. B.	7½	Nov. 28, 1878	Feb. 22, 1883	No.	5	Examined after return from India.
38	E. W. B.	50	Oct. 22, 1879	June 10, 1884	...	5	Very large scrotal.
39	E. P.	3	June 2, 1877	June 19, 1882	No.	5	India; examined after return.
40	Lieut. H. L. C. L.	30	Mar. 12, 1878	Sept. 24, 1883	Riding tr.	5	
41	J. H. L.	4	June 22, 1871	Oct. 20, 1875	No.	4¾	
42	J. D.	1½	Aug. 21, 1862	Oct. 1866	No.	4	
43	J. A. T.	36	Oct. 5, 1872	Feb. 24, 1876	Light tr.	4	Very large scrotal.
44	E. B.	5	Dec. 20, 1873	Jan. 14, 1877	No.	4	Ditto.
45	E. B.	12	Nov. 28, 1863	May 8, 1867	No.	4	Female, right inguinal, congenital.
46	F. B.	10	April 15, 1879	June 2, 1883	No.	4	Passed Navy medical examination.
47	G. D.	7	June 24, 1862	June 1866	Light tr.	4	Enormous congenital; operated on three times.
48	M. H.	40	Nov. 9, 1881	Jan. 1885	No.	4	Omentum and sac removed.
49	R. M.	27	Oct. 1, 1880	Mar. 1884	No.	4	Irreducible; omentum and sac removed; large.
50	C. D.	4½	June 28, 1862	June 1866	Light tr.	4	Operated on twice; very large congenital.
51	M. L.	34	Feb. 28, 1881	Jan. 1885	Light tr.	4	
52	Capt. C.	35	Mar. 20, 1864	April 1, 1867	Light tr.	3	Ruptured other side by fall from horse, hunting.
53	A. B.	30	Aug. 6, 1881	June 1884	No.	3	Sac and omentum removed.
54	J. C.	18	April 8, 1867	May 1870	...	3	
55	M. E.	18	Jan. 12, 1881	July 24, 1884	No.	3	
56	P. S.	8	Nov. 10, 1880	Oct. 1883	No.	3	Passed Army Medical Board.
57	J. D. R.	1½	Aug. 5, 1881	Dec. 3, 1884	...	3	
58	C. K.	8	April 1876	June 20, 1879	No.	3	
59	J. S.	22	May 18, 1872	June 30, 1875	No.	3	
60	W. S.	22	Mar. 29, 1862	July 9, 1865	No.	3	
61	G. J.	18	Dec. 1, 1869	Dec. 1872	No.	3	
62	M. D.	18	Jan. 6, 1874	Jan. 1877	...	3	

No.	Initials	Age	Date of Operation	Date last seen	Truss	Years	Remarks
63	C. D.	4½	Jan. 28, 1862	Oct. 1866	...	3	Heard of.
64	S. B.	26	Feb. 27, 1864	May 26, 1876	No.	3	
65	R. H.	30	July 12, 1862	June 1865	No.	3	
66	D. S.	21	July 2, 1865	May 1868	.	3	
67	J. P.	19	Dec. 30, 1865	July 10, 1868	...	3	
68	I. A.	19	Nov. 1, 1862	Dec. 1865	...	3	
69	W. B.	19	Nov. 10, 1880	Dec. 1883	No.	3	
70	M. E.	18	Jan. 12, 1881	July 24, 1884	No.	3	
71	P. S.	8	Nov. 10, 1880	Oct. 1883	...	3	
72	J. W. R.	1½	Aug. 5, 1881	Dec. 3, 1884	...	3	Very large congenital scrotal; uncontrollable.
73	A. J.	13	Feb. 19, 1882	Nov. 21, 1884	No.	2½	Undescended testis put into scrotum.
74	I. P.	18	Oct. 19, 1861	Jan. 1863	...	2	
75	U. B.	37	July 9, 1881	Nov. 1884	...	2	Passed Medical Examination for India.
76	M. B.	28	May 1, 1882	June 1884	...	2	Cornet player.
77	M. C.	20	Nov. 12, 1874	Dec. 1876	...	2	
78	J. E.	27	April 30, 1881	Mar. 18, 1883	No.	2	
79	G. H.	9	July 4, 1871	Oct. 1873	No.	2	
80	A. S. H.	35	May 9, 1879	Dec. 1881	No.	2	Heard of as quite cured.
81	H. H.	18	Dec. 7, 1881	Nov. 1883	...	2	
82	A. G.	6	July 31, 1883	Feb. 10, 1885	...	2	
83	J. K.	10	April 19, 1883	Jan. 5, 1885	...	2	Other side weak; wears light double truss.
84	A. L.	18	Dec. 16, 1882	Oct. 13, 1884	...	2	
85	J. E. M.	21	April 3, 1882	June 1, 1884	No.	2	Passed Army medical examination.
86	Mrs. P.	30	Oct. 5, 1878	June 21, 1880	No.	2	Mother of child also operated on.
87	A. C. R.	26	Mar. 20, 1880	May 1882	No.	2	Heard of from another patient sent by him.
88	K. R.	10	Mar. 19, 1882	May 20, 1884	Light tr.	2	
89	L. S.	2½	Oct. 19, 1861	Oct. 2, 1883	No.	2	Very large congenital.
90	G. S.	20	July 11, 1883	Mar. 1885	No.	2	Slight bulge after operation.
91	J. W.	20	May 20, 1880	Dec. 1883	...	2	
92	J. W.	7	April 6, 1880	Feb. 3, 1883	No.	2	
93	A. W.	22	Feb. 15, 1881	Oct. 9, 1883	No.	2	Large congenital.
94	A. B.	30	Nov. 25, 1876	Oct. 15, 1878	...	2	
95	W. A.	35	Oct. 5, 1872	Jan. 1874	...	2	
96	C. S.	21	May 18, 1878	Nov. 7, 1880	...	2	

E

operation and of the time of the latest examination of the patients, or of hearing of them from competent authority, are given in the fourth and fifth columns.

It should be mentioned that No. 49 in the table was a case in which the omentum, though not adherent, could not be returned ; while No. 73 was complicated with undescended testes. Both were wire cases, in which the sac was not separately ligatured.

It may also be pointed out that No. 6, after seventeen years without truss, by falling downstairs with a load of bacon on his back, ruptured himself again on the same side. This can scarcely be considered in any other light than as a fresh and distinct rupture, which would have occurred even if the side had never before been ruptured. Nos. 26 and 27 were double ruptures, both operated on successfully at about the same time, and counting really as two more successful operations to be added to the ninety-six, making ninety-eight in all. Nos. 10 and 47 were enormous uncontrollable ruptures, each operated on three times before. a completely successful result ensued, the size of the rupture being diminished by each operation. No. 47 was in a child of seven years, and the return was mainly caused by violent crying. Five cases have been, since the cure, in service in India, the relaxing climate of which is highly unfavourable to rupture cases. Three cases passed, after the operation, the medical examination for the army and one for the navy, after being before rejected for rupture.

The ninety-six cases thus tabulated are those in which the operation therefore was found to be

permanently or durably successful, and entitled in
every sense to be considered as "radical cures."
Out of the total number of cases of the same class,
viz. 339, there are 152 others which have been
examined at periods under two years from the time
of the respective operations, and found satisfactory.
Fifty-nine were found on after-examination to be more
or less failures in the intention of producing a radical
cure. Most of these, however, were improved by the
operation, and a truss was made available which had
not been so before. None of them were made worse
by the operation. The remaining cases have not
been seen or heard of since their discharge from the
hospital.

The number of successful operations out of a total
of 339 (those of above and under two years' duration
at the time of last examination taken together, as
I think may fairly be done) is therefore 248, while
the number of ascertained failures is fifty-nine.
This result gives about 73 per cent. of successful
cases ; and it is reasonable to suppose that the same
proportion may be maintained in the cases of which
the result is unknown at the present time.

If we compared the nine-six cases, or ninety-eight
operations of known successful results, with the fifty-
nine cases of ascertained failures, in order to obtain
our percentage, by submitting to this we should still
obtain nearly two-thirds of successful cases ; and this
notwithstanding that the ninety-six cases, or ninety-
eight operations, are only reckoned from those of
over two years' standing, while the fifty-nine cases
chiefly became unsuccessful before the end of the
first or in the second year, during which period the
majority of the failures declared themselves. If from

the total 339 we deduct the earlier cases of thread
and compress and pin operations (73 in number),
and take only the results of the improved operation,
we obtain 82 per cent of successful cases.

Spray and Antiseptic Cases.—In twenty-eight
cases of inguinal hernia the operation was conducted
under the spray, with antiseptic gauze-dressings.
The sac was tied at the neck with separate stout
catgut, and removed ; while the sides of the canal
and rings were drawn together by kangaroo- or ox-
tendon, or wire.   All the cases were large and
severe.   Sixteen were reducible ; and of these one
died of broncho-pneumonia, with some *post-mortem*
signs of septic infection.   The man was a marine
engineer, aged twenty-six, who had served long on
the coast of Africa, and had fever several times.
He was a bad subject for operation, but was completely incapacitated for his work and livelihood by
a large and uncontrollable rupture.   Twelve were
cases of irreducible hernia from adhesions, &c., all
severe, many very painful and tender.   Of these,
two died.   One, aged forty-five, was operated on
during a very foggy and severe winter, and died
of broncho-pneumonia.   The rupture was large,
adherent, and painful.   He begged earnestly for the
operation to be done for his relief.   The other fatal
case was complicated with a cystic growth on the
spermatic cord and epididymis, which was removed
with the sac.   His age was fifty-six, and he had
come from the Cape to have the operation done.
He was warned of its severe nature, but had suffered
so much from its unmanageable nature and misfitting
trusses that he was incapacitated from both work
and enjoyment.   He died from pneumonic conges-

tion and bronchitis. No failures to cure have hitherto resulted from this operation ; but the percentage of fatal cases is considerably increased, in comparison with the more strictly subcutaneous method, in which the sac was not removed. No doubt this is partly the consequence of the greater severity of the cases, as well as of the operation. Three deaths out of twenty-eight cases gives about 11 per cent. of fatal cases, as compared with $1\frac{1}{2}$ to 2 per cent. Four males and two females have been operated on more than two years, and have shown on examination no signs of a return. In six other cases (males) the results have not been tested or verified since the operation. Twelve males and three females have been cured for a period not yet reaching two years.

The mortality agrees fairly with that obtained by Tilanus of Amsterdam—namely, about 11 per cent. —in his collection of one hundred continental cases of the radical cure of hernia by the open method of operation by dissection, with and without the use of the more strict antiseptic precautions. If this death-rate be found to be maintained on further experience, it seems somewhat too great a general risk for an operation of the class to which that for the radical cure of hernia belongs—namely, that of expediency. In the collection of cases made by Dr. Israelsohn, and given by Professor Annandale of Edinburgh, out of seventy-one cases, four of the patients died, while sixty-six were cured, the operations being performed by various surgeons, but all with strict antiseptic precautions. This points to a further improvement to be obtained by the careful carrying out of antiseptic precautions.

Hernia complicated with Retained Testis.
—These cases are invariably of the congenital
variety of hernia, and are often accompanied by
other morbid conditions, such as adhesions and
intrasaccular obstructions or strangulations. When
orchitis or epididymitis has been present, which is
not unfrequently the case, especially if a truss have
been worn or injury sustained, very difficult compli-
cations may ensue. In two cases the testicle was
retained in the abdomen, and could not be felt or
reached by opening up the canal. The patients
were young adults, who had worn trusses for some
years, with much pain and discomfort, and without
effect in keeping up the rupture. When the
testicle was drawn down by the descent of the bowel
(to which it was in all probability adherent in the
iliac fossa) the truss could not be worn at all, and
much inconvenience ensued from the presence of the
testis in the deep ring. Under these circumstances
the canal was closed up to the deep ring by the wire
suture, the neck and fundus of the sac being tied
and removed. One of these was operated on a
second time, the first operation having failed. The
ultimate result was the comfortable wearing of a
light truss. In another case the hernia became
acutely strangulated, and necessitated immediate
operation. On opening the canal under spray there
were found two deep hernial openings, one placed
internally and filled up by protruded and adherent
omentum, and the other, externally, was occupied by
a knuckle of strangulated intestine, lying along and
over the cord, and adherent to a very small and
shrivelled testicle. The testicle was removed, the
spermatic artery tied with catgut, and the omentum
and double-necked sac also tied separately and re-

moved flush with the peritoneum. The canal and superficial ring were then wired up with a small drainage tube placed along it. The patient did extremely well under the gauze dressing, and recovered in a short time with a bulgy groin, produced by that deficiency in the development of the internal oblique and transversalis muscles to which I have before alluded. He wears now a light truss for supporting the weak parts.

In six other cases the testicle was by an open incision freed from its adhesions and abnormal attachments in the canal and rings. In all of them the cremaster muscular fibres were wasted or absent, and the adhesions were united with its connecting fascia and the conjoined tendon, and with the fascia propria, and in two cases with the inter-columnar fascia. In these two last cases the testicle lay in the superficial ring upon the crest of the os pubis. In all of these six cases the congenital sac was formed, as is usual, by the tunica vaginalis, which, with the globus minor of the epididymis, was drawn down below and in front of the testicle. In three of them the epididymis was spread out and drawn down below the testis in a manner illustrative of the action of the gubernaculum, which, attached mainly to the epididymis and peritoneum, caused these structures to precede the adherent and delayed testicle in its descent.

By separating the adhesions carefully and removing the adherent omentum when necessary, the testicle could be cleared, and the spermatic cord examined and freed from adhesions. The cord was then carefully stretched by being pulled forwards and outwards, and then slipped down into the upper contracted part of the imperfect scrotum, previously dilated by the introduction of the finger, and freely stretched like a glove until it was large enough to hold the testicle. A thick silk or tendon ligature was then passed through the hinder and lower part of the scrotum by means of an ordinary curved suture needle, or a handled needle. Then it was made to pass through the fibrous tissues in close contact with the testicle and spermatic duct, and out again through the scrotum, about an inch distant from the first puncture. The ends of the ligature were then tied over a carbolized pad of the size of the end of the thumb, at the bottom and back part of the scrotum.

In three cases the spermatic cord was found to be too short and resisting to be placed, without great tension, in the scrotum. To overcome this difficulty I carefully dissected with the point of the scalpel through the connective tissue attaching the vas deferens to the globus major, as far down as to enable me to turn the testicle upside down, with the lower part of the epididymis and globus minor still attached to the testicle. By this means I gained the length of the testicle (about one and a half inch) which without further strain lay topsy-turvy in the scrotum, the cord and epididymis being above it. A drainage tube through the bottom of the scrotum, and the use of the spray and careful dressing with gauze, resulted in perfect success.

In one or two cases the testicle showed a disposition to ascend as far as the root of the scrotum by subsequent contraction, but remained placed out of the way of injury below the penis, between its root and the origin of the adductor longus muscle. In none of these cases has the testicle been found to waste since the operation, and this must be attributed to the care with which the deferent vessels, passing from the testicle to the epididymis, and the numerous small branches of the spermatic vessels proper, were arranged and disposed.

I have transplanted the testicle in this way in six cases in which no hernia was apparent. In one of them, a little child, the testis, after a few months' interval, was not to be found, and had either atrophied or reascended into the canal. In the cases in which the upper opening of the scrotum was wide a stout tendon or catgut buried suture, placed across it when the testicle was lodged in the scrotum, assisted in keeping it there. This could usually be felt for many months after the operation as a distinct ridge above the testicle.

In two cases a hernia appeared in the inguinal canal after the testicle had been transplanted for some months. These had either been overlooked at the time of the operation, or had formed subsequently. In one, the operation for the radical cure was subsequently performed, and the other still wears a truss, with a pad of the shape of a horse-shoe pressing on the canal, the breadth of the pubic bone intervening between it and the transplanted testis. Since these cases occurred I have made it a rule, if the canal be patulous when the testis is transplanted, to put tendon sutures to its sides and rings, as in the subcutaneous operation for the radical cure, considering the great probability that a hernia will follow the testicle along the funicular process into the hollow made by it during its retention in the canal.

In all these cases the stretching and the detachment of the vas deferens from the testicle and epididymis were much facilitated by the loose unravelment—a spreading out which the epididymis and tissues had undergone by the continuous traction of the gubernaculum since the fœtal period. In several the epididymis was pulled to a considerable distance from the testicle.

In the cases in which the open and dilated funicular process, which forms the sac in congenital hernia, was long enough to permit it to be brought down into the scrotum, the fundus was utilized to form a tunica vaginalis, by being detached from the neck of the sac, and stitched up with a glover's suture of thin catgut, on a level with the top of the testicle. The intervening neck of the sac, from this point to the deep ring, was then detached, tied with a double ligature of tendon or catgut, flush with the deep. hernial opening, and removed altogether. In some of the cases a twist or two was given to the neck of the sac before the ligature was applied and the sac removed. In one case, where the gland was evidently atrophied, flabby, and little more than fibrous connective tissue, the testicle was entirely removed. With this exception, and the case of strangulation before described, the testis was preserved, and the result as to the rupture was even more satisfactory than where the gland was removed.

LECTURE III.

Anatomy of Crural Hernia.—In the hollow of the groin, below the inner extremity of Poupart's ligament, and forming, with the deep and superficial abdominal rings, an inverted triangle, of which it is the apex, lies the saphenous opening.

Formed by the separation of the fascia lata into two layers, of which the outer or iliac part is advanced forwards by its attachment along the whole length of Poupart's ligament, and the inner or pubic portion, covering the sloping surface of the adductor and pectineus muscles, is attached to the pectineal line, and continued behind the sheath of the femoral vessels to be continuous with the iliopsoas fascia, and connected with the capsule of the hip-joint; the saphenous opening presents an oval outline, looking forwards and a little downwards and inwards. The outer border curves sharply inwards in a falciform manner, lies in front of the femoral vessels, and is attached to and blended with Gimbernat's ligament by a process about half an inch wide—Hey's ligament, important to surgeons, because it crosses the upper part of the crural canal, and may be the seat of strangulation. Below, the falciform edge passes under the saphena vein, which curves

over it to join the common femoral vein, receiving
as it does so the veins which accompany the superficial branches of the common femoral artery, namely,
the two external pubic, the epigastric, and the cir-
cumflex iliac.   The first two pass inwards, the next

FIG. 16.

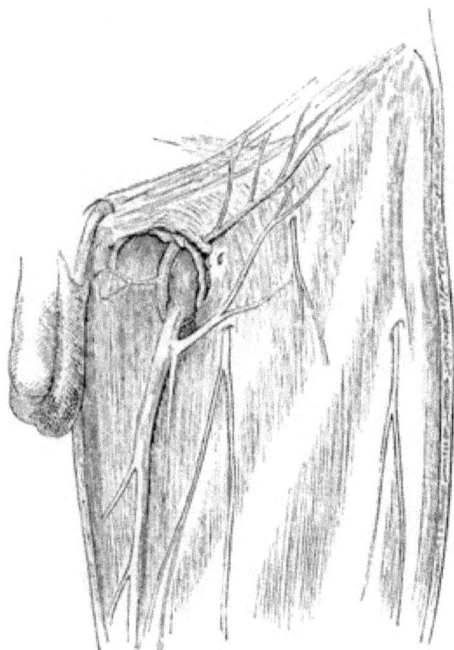

Superficial Dissection of the parts of Crural Hernia : showing
Saphenous Opening and Fascia Propria.

upwards, and the last outwards.   They supply the
numerous lymphatic glands which lie in two groups
in this situation, the upper directed obliquely along
Poupart's ligament above and invested by Scarpa's
fascia, and the lower lying parallel and internal to
the saphena vein.   Nearly all the afferent ducts of

these glands pass in a body through the saphenous opening, the inner part of the femoral sheath, and the crural ring, to join the deep iliac and lumbar glands. The saphenous opening is covered in by a layer of fascia connected with its borders, called, from the numerous holes which transmit these

FIG. 17.

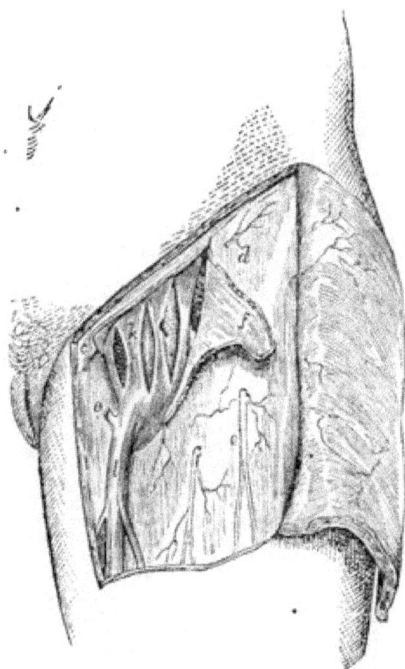

Dissection of parts of Crural Hernia : showing the three compartments of the Femoral Sheath.

lymphatics and some of the vessels, the cribriform fascia. This is blended, superficially, with the deep layer of superficial fascia of the thigh, and deeply with the inner part of the sheath of the femoral vessels.

By detaching Hey's ligament from Poupart's, and turning down the fascia lata and cribriform fascia,

the sheath of the femoral vessels is brought to view.
It is arranged in three compartments : the outer for
the common femoral artery, the middle one for the
vein, and the inner one, smaller and funnel-shaped,
is filled up by the afferent lymphatic ducts. On
clearing these, it is found that they are held in
position by a thin horizontal layer of perforated fascia
containing a lymphatic gland, derived from and continuous with the subperitoneal fascia and fat of the
iliac fossa, attached externally to the strong longitu-
dinal septum covering and protecting the femoral
vein, and internally to Gimbernat's ligaments.   This
is the *septum crurale* of Cloquet, and is of little or
no surgical importance.

If the finger be pushed upward into the iliac fossa,
it will pass through the crural ring.   Arching over
it, in front, will be felt a curved border, formed by
the union of the fascia transversalis with the deep
fibres of Poupart's ligament, the deep crural arch.
This is the usual seat of strangulation in crural
hernia. ·Outside is the femoral vein, covered by its
septum ; inside, the edge of Gimbernat's ligament, to
which the deep crural arch is attached ; and behind
is the hinder part of. the crural sheath, resting upon
the pectineal line of the pubis, and the pectinis muscle
arising from it, and covered by the strong pubic
portion of fascia lata.   Above and externally the
epigastric artery arises from the external iliac trunk.
Sometimes this gives origin to an irregular obturator
artery, which may then descend into the pelvis close
behind the border of Gimbernat's ligament.

Sometimes this branch arises from the external iliac
itself, and then it passes down external to the crural
ring altogether, and the precautions taken to avoid

injuring the iliac vessels will also avoid the irregular branch. It is but rarely, however, that this vessel is cut in relieving the strangulation of a crural hernia. It yields readily to pressure in the loose subperitoneal investment. If the hernia knife be not too sharp, it will cut the tense fibres of Gimbernat's ligament without injuring the loose, yielding and elastic artery. In dividing the strangulating portion of the crural ring the knife is usually directed inwards.

On opening the peritoneum and dissecting the crural ring from the abdominal cavity, it will be seen that a depression is formed opposite to that opening, internal to the iliac vessels. When the os pubis presents a salient border or projection at the pectineal line, as in old age, it becomes more easy for the bowel, by its pressure, to obtain a purchase or hold for the formation of a hernial sac. It will also be seen that the peritoneum around is more closely attached to the subjacent structure than in the inguinal region. Hence the sac of a crural hernia is rarely so large as that of an inguinal one. The aspect of the crural ring is more horizontally placed than the deep inguinal ring. It faces upwards and downwards, with a slightly forward direction. The best guide to it is the pubic spinous process, which is about half an inch internal to and a little above it ; while Poupart's ligament is directly above.

The most frequent seat of strangulation is at the deep crural arch, or in the neck of the sac, at that point thickened by the continued pressure. Next frequently, the strangulating point may be the edge of Hey's femoral ligament, when the sac and bowel are bent outwards at an acute angle. Ill-directed attempts at taxis may here do much harm, and may

even cause ulceration of the bowel at the point
of constriction. This is especially apt to occur

FIG. 18.

Diagram showing shape of Crural Canal on left side,
and a complete Crural Hernia on right.

in elderly females, whose tissues are thinned and
delicate, and the edges of the fasciæ and pectineal
line sharp.

In a large crural hernia the manipulation of the
taxis should be at first directed inwards and downwards, and then backward, with a slight upward
inclination of the neck of the sac. It is rare that
directly upward pressure is required.

Formation of a Crural Hernia.—First, the
sac is pressed downwards and slightly forwards,
dilating the canal, along the lymphatic compartment
of the femoral sheath, pushing before it the septum
crurale, and spreading out the lymphatic ducts. It
is then directed forwards, pushing before it the front
wall of the innermost compartment of the femoral
sheath and the cribriform fascia, and forming a
tumour in the groin below Poupart's ligament and
inside the femoral vessels. It then passes outwards
and a little upwards, across the femoral vessels,
where the fascial connections of the falciform process
are weaker and looser than on the inner side, which
is, moreover, supported to some extent by the
numerous small vessels and lymphatic ducts before

described. Finally, it lies with its fundus placed upon and above Poupart's ligament under the integuments, and its neck considerably in front of and close to the femoral vein.

In the course of the formation of a large hernia Poupart's ligament is bulged forwards, and the groin

FIG. 19.

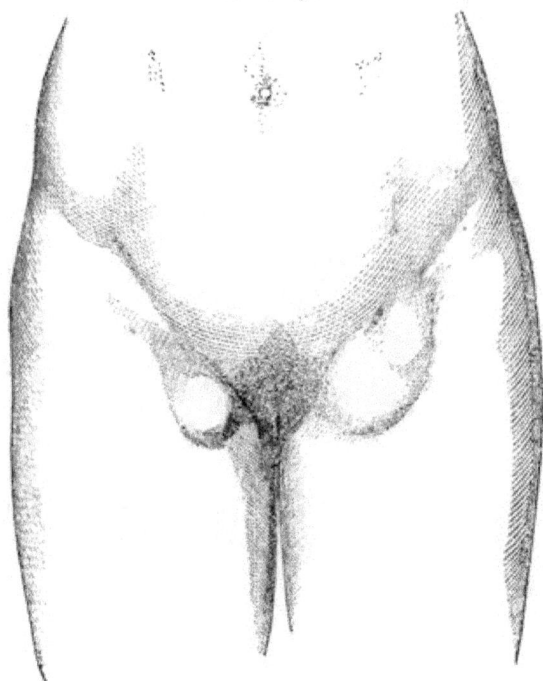

Crural Hernia in the female: small on right and fully developed on the left side.

loses more or less of the hollowness that it presents normally over the saphenous opening. The greater length and slenderness of Poupart's ligament in females, with the greater width of the pelvis and of the hips, render them more liable to this form of

F

rupture than the male sex ; while the latter, from the presence of the spermatic chord, are more liable to open abdominal rings are inguinal hernia.

The coverings of a crural hernia are :—1. The integuments, often thickened by the presence of a deep layer of fat, which may mask the position of

FIG. 20.

Hour-glass constriction of Sac in Crural Hernia by superficial vessels.

Poupart's ligament and render the diagnosis some-what difficult ; enlargement of the femoral glands in these tissues may also tend to obscure the diagnosis, and mask the saphenous opening. 2. The cribriform fascia, recognized by the perforations of the lymphatic ducts, often also obscured by fat and glands. 3. The fascia propria, or crural sheath, recognized by its superior density and fascial appearance. 4. The septum crurale, or subperitoneal fascia, covering the sac, recognized by its greyish-blue, sometimes greenish colour. 5. The sac itself is often thickened, wholly or in part, sometimes presenting an hour-glass shape, from the crossing of vessels, regular or irregular. In one case, the shape was due to an irregular origin of the deep epigastric by a common trunk with the obturator from the common femoral, the former crossing and deeply indenting in front a crural hernial sac, and thus forming a very dangerous irregularity. The crural hernial sac and its contents are more liable to adhesions and abnormal positions of the omentum and bowel than an inguinal hernia.

Thickening of the sac at the neck and adhesive blendings of the covering structures and glands may render it difficult to discriminate the tissues during operation; or a varicose condition of the saphena or femoral vein may give rise to great difficulty, and need careful precautions. By drawing down the sac and exposing Poupart's ligament and the pubic spine clearly, a good guide is obtained in difficult cases. In large and old-standing cases very often no discrimination between the different layers of covering is possible. In operating, the femoral vein should be carefully protected by the finger or a small spatula, and both Poupart's ligament and the pectineal fascia should be kept distinctly in view, as well as the edge of Gimbernat's ligament.

Operation for the Radical Cure of Crural Hernia.—The skin of the groin being shaved and washed with carbolic lotion, and the patient anæsthetised, a fold of skin over the site of the hernial

tumour should be pinched up, and a scalpel or
tenotomy-knife carried with its edge upwards, cutting
towards the surface in a vertical line. This will
usually expose the cribriform fascia. In a stout
subject it is better to dissect down, through the fat, in
successive layers. The deeper tissues are then cut

FIG. 21.

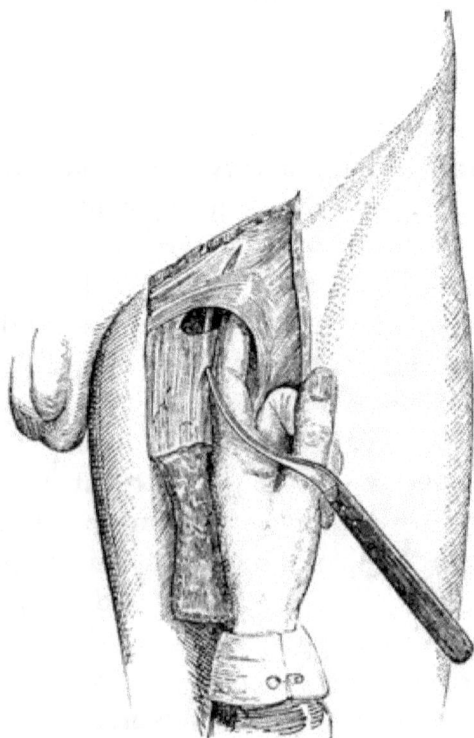

Dissection showing the application of the needle to Poupart's
Ligament in the operation for Radical Cure of Crural Hernia.

through, down to the sac, which is to be opened
by pinching up a portion, lateralizing of the knife,
and, with or without director, slitting it up vertically.
With the handle of the knife the sac is then
carefully separated from the surrounding structures ;

then it is opened, the contents examined, and thickened, elongated, or adherent omentum removed, after ligature of the vessels separately with thin catgut. The sac, being carefully emptied, is to be transfixed at the neck with the handled hernia needle carrying a stout tendon ligature, tied on each side, and cut off close.

FIG. 22.

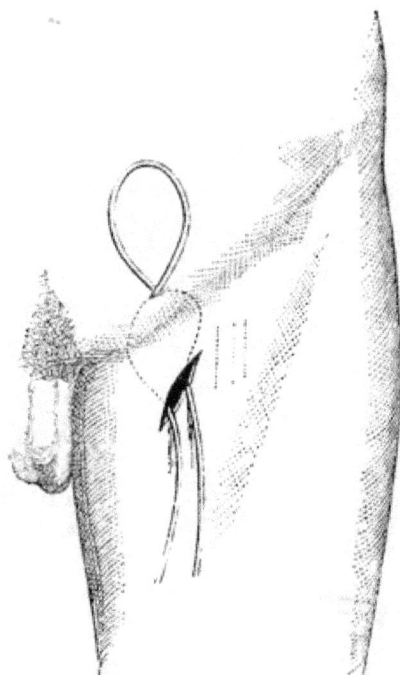

Diagram of wire (or tendon) after application for Radical
Cure of Crural Hernia.

Next the needle is carried through the deep layer of the crural sheath and the pubic portion of fascia lata, entering an inch below the crural ring, and emerging close up to the pectineal line, at the side of the femoral vein, which is to be carefully protected by the finger or a spatula. The needle is then carried through Poupart's ligament, emerging at the upper part of the incision. It is then threaded with one end of the same stout piece of tendon used to ligature the sac, which is then drawn down, emerging at the lower part of the incision. The needle is then again to be passed at the inner side of the wound, through the pubic fascia lata, skirting Gimbernat's ligament, and transfixing a second time the inner end of Poupart's ligament. The other end of the tendon ligature is then attached to the eye of the needle, and brought down, and the two ends are tied with a double surgeon's knot, or, if wire be used, twisted by two turns. The effect will be to draw backwards and downwards to the os pubis the inner end of Poupart's ligament, and to close up the crural ring completely. A piece of small drainage tube, or horsehair, is then placed along the wound, from the ring to the lowest angle ; the wound is closed by closely applied interrupted sutures, and gauze or cotton-wool dressing applied in the usual manner. Latterly, I have found the use of tendon ligature so satisfactory, that for this operation I prefer it to wire. The wound usually closes over it, and heals by adhesion at once, and there is not the pain and inconvenience of the withdrawal of the wire. So far the endurance of the tendon, when buried in the tissues, has been long and satisfactory enough to maintain the cure, which has been watched and noted in some cases for above two years.

Operation for the Radical Cure of Hernia after Kelotomy for the Relief of Strangulation.—At the meeting of the British Medical Association at King's College, London, in August 1873, in the address given by me to the Surgical Section, I described an operation, performed in March of the same year, for the closure of the inguinal canal and rings, after relief of the strangulation by kelotomy. The hernia was a large one ; the bowel was strangulated to a chocolate colour, and the sac contained inflammatory serous effusion. The case did excellently well ; no peritonitis ensued ; and the patient was shown to the members of the Association, wearing no truss. The steps required, in addition to the ordinary operation of kelotomy with opening of the sac, consisted in ligature of the neck, and the application of wire for the closure of the canal and rings, in the same way as in the subcutaneous method. In all cases I have ligatured the neck of the sac flush with the peritoneal opening, and in some removed it altogether, and then laced up the canal and rings with wire or tendon-ligature.

I have operated for the radical cure of strangulated hernia, after kelotomy, in sixteen cases—namely, eight inguinal and eight crural. Seven were done by ligature of the neck by catgut and entire removal of the sac, with the closure of the canal and rings by wire-lacing, and nine by the use of tendon for all these purposes. Of these, one case of inguinal hernia in a male, and one of crural in a female, died from the effects of the strangulation upon the bowel. In neither did the removal of the sac or the closure of the canal appear to add to the risk of a fatal result, inasmuch as none of the *post-mortem* morbid appearances implicated the rings or the canal, but were confined to the bowel, omentum, and visceral layer of peritoneum. Such, indeed, is usually the case in the cases of death from the operation for the cure of hernia. It is rare to find any amount of peritonitis, and still more rare to find morbid appearances affecting the sac or canal.

The history of hernia shows that the sac may be twisted, perforated through and through, injected with irritants, ligatured or removed altogether, or may slough off, without any appreciable amount of peritonitis; and that death, when it ensues, is far more often from blood-poisoning, erysipelas, and such diseases as may occur from any operation, than from peritonitis. We may therefore conclude that, antiseptic means being employed to prevent such results, the attempt should be made to cure the patient of the hernia after kelotomy, in all cases where the condition of the bowel renders it safe to return it into the peritoneal cavity, after the relief of the strangulation. The omentum may be dealt with more freely than has heretofore been done. If it be sphacelated or suspicious in appearance, the vessels may be taken up in a healthy part above, tied with small catgut, and the doubtful part removed by tearing, or with a pair of blunt scissors. This may be done by spreading out the omentum, and taking up the vessels, which are usually plainly seen from the congestion of the veins, with a piece of thin catgut, applied by a common suture needle, an aneurism needle, or a common pair of dissecting forceps.

In a very few cases, indeed, a cure of the hernia may occur after kelotomy, where no attempts have been made to close the sac or canal; but ordinarily the division of the strangulating structures leaves the hernial aperture much more patulous, and the hernia larger and more uncontrollable by trusses than before.

To prevent this by an addition to the ordinary operation of kelotomy, which does not appreciably increase the risk, and is effected while the parts are exposed by that operation, seems to be a course which will commend itself to most surgeons.

Umbilical Hernia. — The anatomy of the umbilicus is sufficiently simple. An opening left in the linea alba at the site of the exit of the omphalo-mesenteric duct, between that and the closed

FIG. 23.

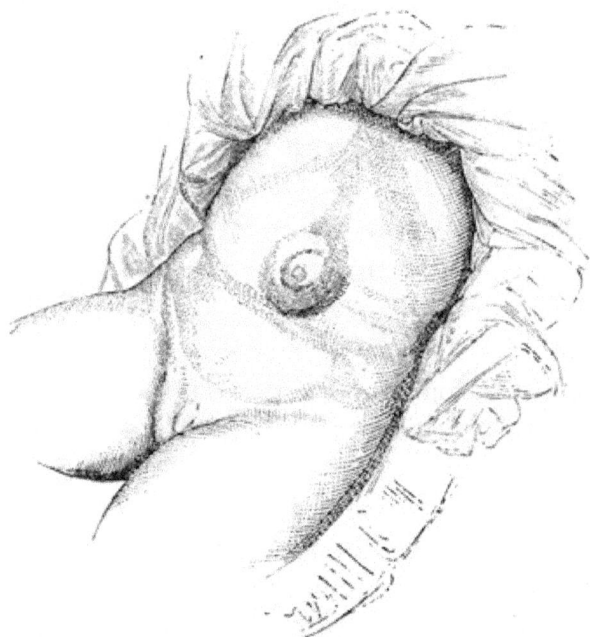

Umbilical Hernia in an infant.

urachus, from imperfect development, sometimes permits at the period of birth a portion of the bowel· to protrude into the substance of the umbilical cord during the pressure and struggles of parturition.

On examining the infant before the application of

the ligature to the navel-string, the appearance of the latter should be carefully noticed when the infant cries. If any swelling take place, or the cord appear thicker than normal, the sac, which is composed of the tissues of the funis, lined by a thin layer of serous membrane, should be carefully compressed, so as to exclude the bowel, if present, and a broad ligature placed flush with the surface of the integuments. By this means the rupture may be at once cured.

To tie the cord at a distance from the navel is in these cases to invite the occurrence of an umbilical hernia ; while no harm can result from the application of a thick silk ligature in the way just described, if the swelling be carefully manipulated to exclude any bowel which may be present in the sac. Often the weakness at the navel opening is unobserved, or the treatment by a pad and belt is discontinued too soon ; and the result is the appearance of an infantile umbilical hernia, when the muscular system of the child becomes more powerful and the efforts in crying more continuous. Sometimes, although no hernia may at once ensue, a weakness is left at the side of the aperture, and a piece of omentum may pass in and out of the sac, preventing the circular contraction of the tendinous cicatrix, which would seal up the opening and cure the hernia. The original weakness may persist, and the first occurrence of the hernia—though this is not commonly the case— may happen in adult life or even in its decline. At this time the accumulation of intra-abdominal fat, the occurrence of pregnancy or ascites, or the habitual distension of the bowels by air, may produce a rupture here. In such cases the navel cicatrix never presents the cup-shaped depression of a strong and healthy development. The peritoneum covering the inner surface of this point, when healthily developed, is tough, greyish-white, and presents small puckerings, produced by the contraction of the cicatrix, which binds the margins of the umbilical aperture together. In hernial cases this part is looser, more bulgy, and less firmly adherent. As the hernia becomes larger by the intrusion of omentum, the peritoneum becomes stretched, and finally so much attenuated as to be incapable of demonstration, so that we may look in vain in a long-established case for a real peritoneal sac. The sac is a blending of the stretched and attenuated cicatrix with the remains of the peritoneal structure.

The abdominal wall gives way, in fact, at its weakest part, under the distension. In such cases that part lies in the navel, and not in the groins. In some instances the small apertures in the middle line aponeurosis, which usually transmit the small intercostal vessels and nerves, may become dilated by the growth of pellets of fat, which gradually enlarge them, and finally may · permit (when the patient from any cause becomes afterwards thin) the occurrence of a rupture close to the navel, and apparently (at first sight) umbilical. It will be found, on close examination of such cases, that the aperture is on one side of the real navel. The umbilical hernia of adults is always formed in one of the ways just described, but most frequently in the way first mentioned, and usually there is a history of an infantile or congenital weakness in the part. From the exposed and prominent position of the rupture the part is liable to the results of pressure and friction or injury. The sac may be rendered irregular in shape, bulging out on all sides like a mushroom, and thickened irregularly by hypertrophy. The omentum within is very liable to become adherent, or thickened in lumps of fibrinous effusion ; or the bowel may be adherent to the omentum or sac.

Such changes are more commonly found at the lower part of the sac. This is brought about by the weight and pressure of the contents downwards upon the sharp lower curved edge of the hernial opening ; while the pressure of the clothing, or even of a belt or truss worn to support the rupture, but often ill-fitting, pressing downwards instead of lifting upward, and uncomfortable to the patient as well as injurious, produces a constant irritation and shifting of the pressure.

Hence it is found that the strangulating part of an umbilical hernia is placed at the lower edge, with the bowel hanging over it, and sometimes sharply ulcerated through, with extravasation of fæcal matter into the sac. Not uncommonly, however, the strangulating part is intra-saccular, produced by the adhesions or omental openings, and may be during operation difficult to find and to separate from the matted tissues.

The transverse colon is usually involved in an umbilical hernia, but by no means invariably so. Frequently the small intestines are found in the sac, generally the omentum ; in a few instances, a portion of the stomach and spleen.

In applying the taxis or any supporting apparatus to a large umbilical hernia, the pendulous position of the fundus must be carefully borne in mind. The pressure should be a lifting pressure towards the umbilical cicatrix. If this be not attended to, the bruised and damaged bowel is pressed and rubbed upon the cutting lower edge of the hernial aperture, and still further damaged, or even ruptured. The fundus should be placed in the hollow of the hand, and lifted up gently, while the fingers of the other hand manipulate the neck of the sac by a gentle kneading motion. ·

Operation for the Radical Cure of Umbilical Hernia.—As a rule, a congenital umbilical hernia, more even than other varieties, has a strong tendency to get well by the progressive contraction of the ring-like cicatrix around. In this it follows the law of all circular cicatrices, which, unless prevented by other conditions, tend to contract towards the centre, and close up the aperture. Time should therefore in all cases be allowed for this beneficent action of the *vis medicatrix naturæ* to exert its influence, favoured as far as possible by properly fitting pad and belt. But in some, either from early weakness of constitution, hereditary predisposition, frequent crying, or the neglect of proper support, the rupture remains uncured up to the adult period, or becomes so large as to be dangerous and unsightly. This constitutes a serious danger in both sexes : in the male, from the disability and weakness thereby ensuing ; and in the female, from the liability to pregnancy.

In only two cases have I thought it advisable to interfere by operation for the cure of umbilical hernia in adult males. One was a finely grown young man, aged twenty-two, with a protrusion of the size of a small apple. He much wished to

serve in the army, but was refused admission by
the medical examiners. He was operated on,
June 16, 1883, made a good recovery, had no unpleasant symptoms at all, and was finally admitted
into the service, where he still remains, cured. The
other case was about the size of a walnut, in a lad
aged fourteen, who had been rejected by the medical
examiners for the navy. The cure was as perfect
as in the last case, and no symptoms worth men-
tioning occurred.

FIG. 24.

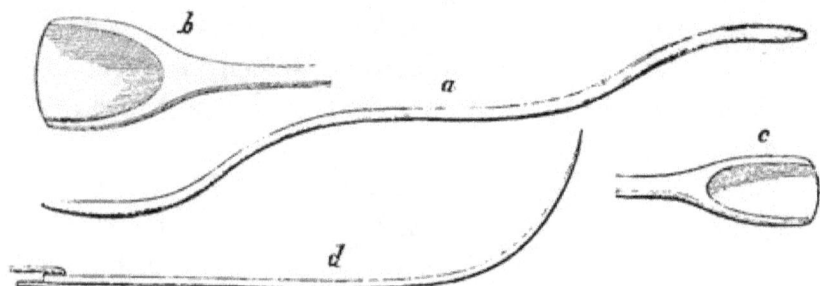

Needle and spoon director for Radical Cure of Umbilical Hernia.

In both cases the operation was done with wire,
in the following manner:—A small semicircular
needle, with a handle, and the eye near the point,
and sharp enough to pierce easily the very tough
and resisting cicatricial tissue around the neck of the
sac, and copper wire silvered, and rather thinner
and more flexible than that used for inguinal hernia,
were the only instruments used in the operation.*
The sac was first pinched up with the thumb and
fingers, and carefully emptied of its contents. It was
then invaginated upon the forefinger, which was
placed under the edge of the hernial opening on one

* In small cases a director, such as is represented in the woodcuts,
will be found necessary to be used instead of the finger for invagination.

side. The point of the needle was then made to
transfix the tendinous margin about half an inch

FIG. 25.

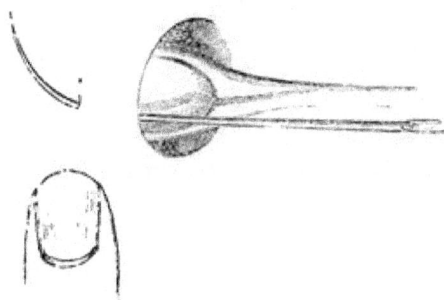

First application of needle in operation for Radical Cure of
Umbilical Hernia.

from the edge. The point of the needle, on raising
and before perforating the skin, was then carried

FIG. 26.

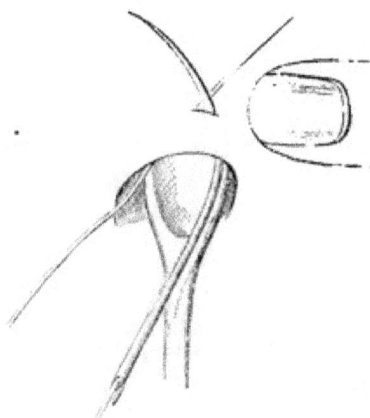

Second application of needle in operation for Radical Cure of
Umbilical Hernia.

round a quadrant of the circular opening, and made
to emerge at the upper pole of the vertical diameter.

The bent end of the wire was then hooked on, and drawn back with the needle. The same manœuvre was then gone through on the opposite side, the upper end of the wire hooked on, and drawn through. The invaginating forefinger was then carried down to the lower pole of the vertical diameter, the needle passed through from behind, and then turned under the skin, so as to be made to emerge where the wire

FIG. 27.

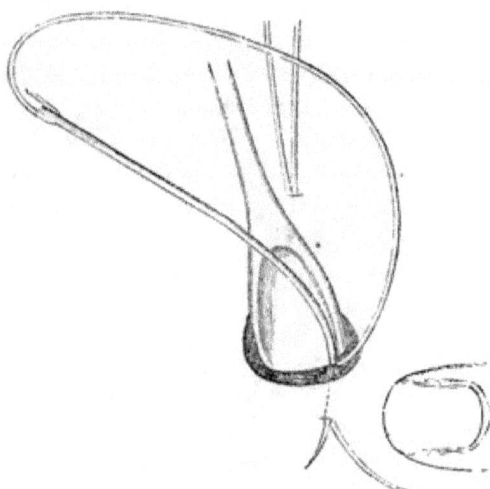

Final passage of needle and ligature in Radical Cure
of Umbilical Hernia.

came out through the lateral puncture. The wire was again hooked on, and drawn down, a loop being left so as to emerge at each of the punctures. The same manœuvre being accomplished at the opposite side, the ends of the wire were left emerging at the lowest puncture, while a loop of wire was found at all the other punctures. The ends of the wire and the upper loop were then drawn upon, until the lateral loops sank under the skin. The invaginated skin

and sac were drawn out of the aperture by traction
with a pair of hooked forceps.   Then, by twisting
down into the punctures the upper loop and the
ends of the wire below, the sides of the hernial opening were drawn powerfully together in a vertical line
coinciding with the linea alba.

The application of tendon may be accomplished
in the same manner, the ends of the ligature being
tied firmly below at the lowest puncture, while the
loops are all sunk into the other punctures, so as to
disappear in the tissues.   The method of application

FIG. 28.

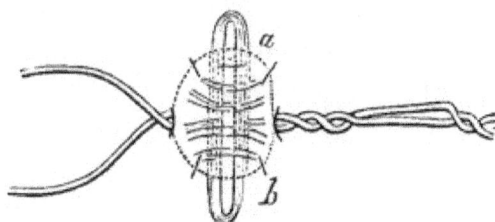

Wire twisted up and neck of sac compressed in cure of
Umbilical Hernia.

of the constricting ligature was in fact almost exactly
the same as the subcutaneous ligature for nævus,
which I have used for many years.

The wire was left in for a fortnight ; much thickening ensued along its track.   It was finally withdrawn by untwisting the loop and lower ends,
cutting off the latter short, and drawing upwards the
former.   The wire, on being straightened by stretching, came out easily.

If the rupture be a large one, it would be better
to keep all the loops, the lateral ones as well as the
upper ones, and twist them all down into their

G

respective punctures. The prepared thick tendon-ligature may be used in this operation instead of wire. It is durable enough, if properly prepared, to endure until it becomes organized, and does not require removal. After the operation the sac of the hernia projects as a vertical ridge, which can afterwards be removed, if desirable, with a pair of scissors, when the cavity is permanently closed at the neck. A pad of gauze, placed vertically on each side, with another thick one over them, covered with oilskin, and held on by a broad body bandage, is all the dressing required.

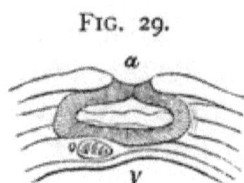

FIG. 29.

Diagram of cicatrix of the track of ligature in Umbilical Hernia.

In three other uncontrollable cases in children I have operated without any serious symptoms. One was by the use of the angular pins locked together. The result in this case was not satisfactory : the rupture returned. The others were operated on by the use of the thinner wire : they were completely successful. No death, or indeed any serious symptom, ensued. The hernial opening into the peritoneal cavity was in each case closed by adhesive effusion, and the wounds scarcely even suppurated. But in the unsuccessful case referred to the adhesions were not strong enough to prevent a return of the rupture.

Review of the Various Methods of Attempting a Radical Cure of Hernia.— I can only briefly allude to the numerous fellow-workers who have followed this line of surgical im-provement. Scarcely any subject in the whole department of surgery has been more discussed or more written about : not any has been more soiled by the practices of mountebanks and charlatans from the earliest dawn of the history of medicine.

The want of permanent success has been almost universal, and the danger to life of many of these proceedings too often demonstrated. Of the ancient methods many owed their fame to extreme ignorance on the one hand and impudent charlatanry on the other. Plasters and ointments of elecampane, caustics and the hot iron, and especially the application of oil of vitrol, brought fame and money to an impudent quack.

Excision of the sac and its covering, and in many cases of the testicle also, was practised by Celsus. Galen and Paulus Ægineta ligatured the sac at the superficial ring, and tied up also the cord and skin. Centuries after their time this was practised by continental quacks, as mentioned by Dionis and Scultetus. Then was introduced the milder and more scientific proceeding of opening the sac freely and stitching its edges close. This was called the royal stitch, because it qualified the king's lieges for military service. The *punctum aureum*, as described by Ambrose Paré, consisted of passing a golden or leaden wire behind both sac and cord, at the superficial ring, and twisting it down tight enough to close the hernial sac, without stopping the circulation through the testicle. In more modern times, Schmucker and Langenbeck exposed the sac by a free incision at the superficial ring, and ligaturing it without enclosing the cord. The fatality of this method (three in ten cases), and the failure of most attempts to cure in the hands of such men as Anel, Armont, and Petit in France, and Abernethy and Sharp in this country, led to the general professional opinion of the uselessness of such operations to produce a cure, and to their actual condemnation as very dangerous to life. The efforts of surgeons then turned in the direction of truss pressure as a means of radical cure. Richter employed a strong and tight truss, with a hard pad of wood ; and this method was employed by L'Estrange of Dublin, who has been followed by many Irish and American surgeons. The injurious pressure upon the spermatic cord, resulting, it seems to me, from the shape of the pad employed, has been marked in some cases I have seen and heard of. The pain and suffering inflicted during the lengthened period necessary for a cure, and the want of skill and care of the patient in adjusting the pressure, and the difficulty and expense of a continual supply of new trusses, combined to prevent a greater success than 10 or 15 per cent. of uncertain cures by this plan. But, undoubtedly, cases of cure by a skilful and prolonged use of an efficient truss have occurred to myself and many other surgeons. In children's cases especially it is much more frequently effected ; and it might be much more common if nurses and parents were as skilled as surgeons and instrument-makers in adjusting the trusses. But too often, especially in hospital practice, the attempt to cure by this method is hopeless ; and the rupture, when brought for radical cure, is large, uncontrollable, and difficult, requiring sometimes two or even three operations.

A class of operations dealing with the interior of the sac by injection of irritants like iodine and tincture of cantharides, was practised unsuccessfully, and sometimes fatally, by Velpeau and Pancoast. Another, dealing with the sac by introducing solid substances, such as threads or sponge, or gold-beater's skin, with the same object, was practised, with equally bad results, by Schuh, Demas, and Riggs. The method of invagination of the skin of the scrotum into the canal, and endeavouring to plug up the rings permanently, was then brought into vogue by Signoroni and Gerdy. This plan was followed by another, namely, the use of a hard wooden plug, forced along the canal, originating in Wützer, and practised by Rothmund, Sigmund, and Spencer Wells.

In all these methods it was found that after a greater or less period of time the plug, at first apparently satisfactory, gradually made its way, pushed by the recurring hernia, down into the scrotum. In many, excessive suppuration brought on burrowing of pus towards the abdomen, and fatal results. The apparent cures were in a great majority of cases temporary and illusory, and the operation fell into discredit. A revival of the injection method, not into but around the neck of the sac, using astringents such as solution of oak bark, has been practised in America by Heaton, and followed by Warren and Bull.

Since the publication of my treatise on Rupture in 1863 fresh methods of proceeding have been originated. Mr. Spanton has invented a new instrument for uniting by adhesions the pillars of the ring. It is shaped like a corkscrew, and is introduced from the groin above downwards into the scrotum, guided by the forefinger, passed through a scrotal incision, into the canal. It is removed after a week or fortnight, according to the action set up. Mr. Fitzgerald of Melbourne, Australia, laces up the pillars of the superficial ring with a continuous gold wire suture, which he leaves in the tissues, in the hope that it may permanently fulfil its functions. Professor Dowell of Texas sews subcutaneously the pillars of the ring with silver wire, and claims to have done a hundred cases, with sixty permanent cures.

It will be observed that in all of this class of operations, the anterior wall of the inguinal canal and its superficial ring are the only parts really affected by the operation. The hinder wall and the deep ring and neck of the sac are unaffected. The result is, that sooner or later the hernia makes its way behind the adhesions or the permanent wire suture, separating the former, and by constant pressure causing the latter to cut its way slowly through the tissues and become useless. In the case of the latter plans also there is the manifest probability of a truss being inapplicable if the hernia return, or a weakness or bulge remain, requiring another operation for the removal of the wires within a few months. The same result will undoubtedly occur if the operation which goes by my name be imperfectly done, and the pillars of the superficial ring only be sutured. A misapprehension which may easily arise from a want of familiar and practical knowledge of anatomy, has led to the application of my name to operations in which all my precautions were neglected; and also to the employment of a part of my operation only as the basis for a new method. The result has been in many instances a want of success, for which the operator, and not the operation, is really responsible.

Reasons somewhat like the foregoing have led to what is called the open method, or the method by dissection, for the radical cure of hernia. It will be well to state here, that, with the protection of the spray and the careful use of antiseptic dressings, I by no means object to, but have often employed, this method of operating. But the real factors in the production of hernia should be properly and securely dealt with. An operation which has been practised by men of ability and professional position cannot be lightly considered by any one. As present workers in this field of surgery I may mention the names of Professors Sir J. Lister, Annandale of Edinburgh, and Stokes of Dublin ; Sir W. Maccormac ; Macleod and Buchanan of Glasgow, Mitchell Banks of Liverpool, and Charles Steele of Bristol—the first surgeon, I believe, who operated for the radical cure of hernia with all the antiseptic precautions.

Professor Annandale of Edinburgh opens the canal, ties the neck of the sac, and removes it bodily, and then stitches together the margins of the opening. In the *Edinburgh Medical Journal* for December 1880 he published a case of strangulated crural hernia, operated on in January 1872, in which he tied the neck of the sac with catgut, and removed it with some adherent omentum. In a case of irreducible crural hernia, operated on in January 1880, he stitched the margins of the crural ring to the stump of the sac in addition.

Professor Stokes of Dublin opens the sac freely, stitches up the neck, and then, without removing the sac, draws together the canal and pillars of the ring by chromicized catgut, carbolized silk, or silver wire. He considers the removal of the sac to be a risky and unsurgical proceeding. Mr. Mitchell Banks of Liverpool opens the canal freely, ligatures the neck of the sac and divides it, detaches the fundus and removes it, and then sutures up with catgut the pillars of the ring.

Mr. Alexander of Liverpool opens the canal by dissection, and ligatures the neck of the sac with catgut, making a point of tying it so that it shall be flush with the peritoneum internally, so as to leave no digital depression. He then divides the neck of the sac below the ligature, and leaves it in the wound, without suturing the pillars of the ring. In one of the two cases of crural hernia operated on by him in June 1880, and reported in the *Liverpool Medico-Chirurgical Journal*, the fundus of the sac afterwards sloughed out. Sir William Maccormac has followed this plan, I believe, in a good number of cases. Professor Buchanan of Glasgow, in cases of congenital hernia, cuts down to the sac, slits it up longitudinally on each side of the cord, then divides the front part horizontally, rolls up the upper half, and with it plugs the deep ring, turning down the lower half to complete the "tunica vaginalis" above.

Many of these surgeons appear to take little pains to close up the inguinal canal, as distinguished from the superficial ring. At the deep ring, some close the neck of the sac without closing the margins of the fascia that forms the ring proper. In an irreducible hernia it is of course necessary to open up the sac freely, and in many cases the inguinal canal also, in order to remove the sac entire. I accomplish this, if found necessary, by extending my usually limited incision upwards from the scrotum. The entire removal of the sac is always a tedious, severe, and often a difficult, operation, if due care be taken to preserve intact the spermatic cord. It must be accomplished by much dragging and tearing, and separating a great number of vessels. The spermatic cord is sometimes found spread out, and its constituents separated widely apart. In two cases of irreducible direct inguinal hernia I found it placed in front of instead of behind the sac—the vas deferens towards the inner and the spermatic vessels to the outer side. In other large cases I have found the vas deferens projecting before it into the interior wall of the sac, with a sort of mesentery thrown around it. Much care is required in dealing with a large sac for its entire removal, and a good deal of extravasation of blood into the penis and scrotum follows in many cases, with often retention of urine. Banks states that it requires a good deal of "mauling"; while Stokes considers the process as "unsurgical," if not "repulsive and barbarous." It may certainly be considered as a most difficult and prolonged operation, and the increased mortality which follows it shows that its effects on the system are serious. It is only very weighty reasons which would justify its peformance.

With respect to the supposed advantages of the open method, enabling the surgeon to see the parts on which he operates, I have myself found that, after the first cut and the application of the sponge, the parts became so bleared with blood that I was obliged to rely mainly upon the aid of the sense of touch, before I ventured to pass a needle through Poupart's ligament, the conjoined tendon, or the pillars of the ring. My experience is, that this operation can be all done, and has been very fre-

quently done by me, when the sac to be removed is
not very large, through a scrotal incision, two inches
long, reaching up to the superficial ring. The
mobility and elasticity of the integuments is such
that the aperture thus made can be drawn up so as
to lie over the hernial opening, which is itself dilated
and large enough to permit the sac to be followed,
drawn out, detached, tied, and divided close to the
deep ring, without any important division of the
intercolumnar fascia, which necessarily weakens the
abdominal wall.

A fallacy which some operators seem to entertain
is, that by stitching the pillars of the ring only, the
canal is also closed. The layers formed by the walls
of the inguinal canal, and the spermatic groove above
Poupart's ligament, in fact remain as loosely connected with each other when the parts are well
healed as they did before the operation. They are
movable upon each other, and slide and give way
before the soft pressure of the sac to form another
hernia. Abundant material for a fresh sac is found
in the peritoneum forming the false ligaments of the
bladder. The plans for simply tying the neck of the
sac, and either cutting it off or leaving it and the
sides of the canal to heal up as they lie, are simply
resuscitations of the older mediæval plans, about
which the late Sir William Lawrence argued " that
it was futile to close up or cut off the neck of the
sac, while the openings in the abdominal wall remained to admit a fresh sac, which the peritoneum
was apt and ready to furnish."

Conclusion.—It appears indubitable, from the
results of the last twenty or more years' experience
of the radical cure of hernia, that the position of

those surgical writers who have maintained that the radical cure should not be attempted except in the severest cases, is untenable. The operation has given as great relief and exemption from the minor troubles and worry which make life miserable as any operation associated with prolapse, such as hæmorrhoids, and is even more safe. It is certainly quite as much called for, on the score of relief from pain and inconvenience, as most other abdominal operations. Though it may not, like ovariotomy, remove the certainty of a speedy death, and may, like colotomy, be called an operation of convenience or expediency, it often relieves suffering as severe as that for which colotomy is performed, and is attended by far happier results.

The justification of the operation being admitted, it remains to consider what cases are most appropriate for it, and which of the many we have passed in review is most proper and applicable for the cases chosen. The rules I have observed in my own cases have been as follows. The subcutaneous plan has been adopted.

1. In cases of children above five years old, in whom trusses are useless and unavailable, because of neglect, violent coughing and crying, sore groins, rapid increase in the size of the hernia, and interference with micturition.

2. In cases of young adults, or boys under fourteen, whose prospects in life as candidates for the naval, military, or engineering professions, or for colonizing, are seriously impaired by the hernial condition. Such persons may be far from surgical assistance when the exigencies of duty or occupation may produce strangulation, or the breakage of

a truss may leave them defenceless ; they are subject also to increased life-assurance rates, from which the operation, when successful, relieves them.   It should be done in able-bodied working men generally, whose various laborious employments may place them continually in danger of strangulation, and whose strength and usefulness are impaired by the hernia.   The extent of the necessity for a radical cure of rupture, and the patriotic and social motives which demand it, are clearly made manifest by the estimates of the number of recruits and conscripts rejected for this complaint.   Malgaigne states that one in every thirteen Frenchmen is ruptured ; Arnaud, one in every eight.   During the Civil War in the United States, 38,132 were rejected in two years. In this country it is said that one in every twenty males is ruptured.   The bodily ailments and mental worry which this condition and its consequences entail upon this large number of human beings make up a very impressive total of suffering.   And the mortality from it must be also considered.   In 1879, according to the Registrar-General's reports, as given by Mr. Spanton, no fewer than 1,119 deaths occurred from hernia, of which 23·5 per cent. had undergone operation for strangulation, &c.   The average rate of mortality of the operation of kelotomy in eleven large hospitals is given by the same author as 41·8 per cent.   The proportion of the mortality from hernia increases with age to a marked degree.   The importance of a permanent cure effected during youth for so large and useful a class as this, when thus viewed, rises to the point of a national demand.

3. In reducible cases, where the sac is thick and

indurated from truss-pressure, or where the omentum is continually slipping down under the truss, showing thereby that it is abnormally elongated, I open the sac, tie the vessels of the omentum separately, and remove it below the ligatures ; tie up the neck of the sac flush with the peritoneum at the deep hernial opening, and apply wire or tendon ligature to the canal and rings. When from any cause a first operation fails in effecting a satisfactory cure also, I open the sac, inspect its interior to discover any special cause for the failure, tie and remove the sac, and lace up the canal and rings with especial care and security.

4. In all favourable cases of straugulated hernia, both inguinal and crural, the coverings and front wall of the canal being necessarily divided to search for the constricting tissues, I open the sac, examine the contents, remove adhesions and doubtful portions of omentum, then tie up the neck of the sac, cut it off short, and remove it altogether (except in congenital hernia), and secure the walls of the canal and rings, as in the subcutaneous method. Of course, a wrong diagnosis of the condition of the bowel or omentum, and of their fitness to be returned into the abdomen, or some other cause arising from the strangulation, may in these cases result in a fatal issue. But I believe strongly that, if drainage be free and skilfully arranged, no increase of risk ensues from the attempt to produce a radical cure. Quite lately, I have done this in a case of *reductio en bloc* in a man who is now convalescent in the hospital.

5. Cases of irreducible hernia, and of large and unmanageable cases of reducible hernia, in patients

otherwise in a good state of general health, and not above the age of sixty, and in which truss-pressure entirely fails to render the patient comfortable and free from danger, seem to me to justify and to require operation, if the patient wish for the benefits which he may reasonably expect from a carefully conducted operation under strict antiseptic methods. In all cases, he should have the chances fairly laid before him in a way that he can understand, and then have the option without bias or persuasion.

In these cases, as in the last class, the operation necessarily assumes more or less of the character of an open operation under spray. The sac is freely opened, and is tied and removed ; but the suturing of the canal and rings is effected as in the subcutaneous method.

In bringing my allotted task to a conclusion, I am deeply conscious that a very considerable part of these lectures have gone over ground which has been elementary enough to re-echo the household words of every dissecting-room and tutor's class. In extenuation I have to plead that the subject demanded a good deal of such anatomical repetition, to preserve its coherency and to make it understood.

After all my efforts, I fear that there still remains much room for elucidation ; and I have been led to think so because men who have arrived at considerable position in the profession have declared to me that they were unable to comprehend my method of operating until they had seen it done by me, and that it was much more simple to see and to do it than to describe it.

At the risk of being tedious to the more experienced of my hearers, I have endeavoured to be clear to those less learned and less experienced. I have some fear that there is no doubt about the first, but that I may not have succeeded in the last as fully as I could have desired. At any rate, I have done the best that was in me to ascertain the exact truth about the possibility of a permanent cure of hernia, to prove it as convincingly as I was able, and to state it as simply as I could, and as shortly and completely as the scope of these lectures has demanded.

The last part of my duty now presents itself, and that is to thank you, Mr. President and gentlemen, most sincerely for your presence and kindness during my lectures.

[Cases of the radical cure of hernia, of twenty-three years' duration and downwards, were shown at the close of each lecture.]

CLINICAL LECTURE

APPLICATION OF TRUSSES TO HERNIÆ.

GENTLEMEN,—We have lately had in the wards a good number of cases of children with hernia—a very common complaint, and one which may be treated with a considerable amount of success by proper care and mechanical restraint. I propose to speak to-day, therefore, of the application of trusses to hernia. It is one of great practical importance to you, both in the immediate future and in your career as practitioners.

I shall commence with inguinal hernia, which is by far the most common ; and we will consider what you may expect to accomplish by trusses in the way of cure, the kind of truss to be applied, and the sort of apparatus which is either of no use or positively injurious, and is therefore to be avoided. Femoral hernia comes next in point of frequency in the adult ; and lastly, we will consider briefly the truss treatment of umbilical hernia.

The relative positions of the apertures which are concerned in inguinal hernia are to be carefully recognized. Under the term "oblique inguinal herniæ" are included congenital and infantile, because children's herniæ descend along the course of the spermatic cord. In the infantile variety the tunica vaginalis is imperfectly closed, and extends high up the canal, and the rupture-sac comes down within it, so that you have reduplication of the serous sac of the rupture by the addition of the front layer of this serous tissue. Now let us study the hernial apertures. First of all, placed a little inside the midway point between the anterior superior iliac spine and the spine of the pubis, you have an internal or deep ring which transmits the constituents of the cord. This is not a direct opening, which you can see on the superficial as well as on the deep surface, but it is like the opening of the sleeve of your coat in relation to the body, which is seen inside and not outside. If you look at it from the outside you will see the infundibular process of fascia passing from the transversalis, covering the opening by being prolonged over the cord, and forming its intimate investment. In the opening of the coat-sleeve the edge is most prominent on the inner part, where the armpit of the wearer is often galled by it. Now, it is just so in · hernia. That is the point also where the internal abdominal ring galls the protruded bowel, and sometimes strangulates it. We will leave out of consideration altogether the relation of the epigastric and other vessels, which scarcely bear upon the treatment by truss pressure. The inguinal canal slopes downward and inward under the lower borders of the internal oblique and transversalis muscles ; and when the hernia has dilated this canal and arrived at the external ring, it forms an opening placed above and outside the cord. Then the hernia, if permitted to go on, slips under the coverings of the cord, and becomes scrotal hernia. Generally speaking, there is a little constriction in the upper

part of a scrotal hernia, indicating the commencement of the unstriped muscle, or dartos, of the
scrotum, and the termination of the fat of the
tegumentary tissues.   This sometimes gives a complete hour-glass appearance to the sac, such as is not
uncommonly also seen in hydrocele.

There are various degrees of oblique inguinal
hernia.   First of all, when the bowel begins to pass
through the internal ring, it does not usually come
through suddenly.   In very few cases does oblique
inguinal hernia occur suddenly at this point of its
progress.   It is prepared slowly; till at length, on the
patient making an effort, it advances a little further,
so as to give him pain, and draw attention to it.   It
occurs in this wise :—There you will see is a ring,·
and behind the ring is the peritoneum.   The peritoneum there is loose, because it has to provide for
the superior false ligament of the bladder, and to
allow of the distension of that organ.   If the subperitoneal fascia and the fascia transversalis are
feeble, and their support to the peritoneum insuffi-
cient, the latter gets so thinned and weakened as to
yield before a cough or any lifting effort.   Even the
slightest causes constantly repeated will slowly and
often unobservedly produce a protrusion.   At length
the hernia gets into the upper part of the inguinal
canal, when it is called a *bubonocele*—*i.e.*, a hernial
tumour, which is still within the inguinal canal, or
just emerging from the superficial ring.   Even in
this condition it may get strangulated at the neck of
the sac as it passes the internal ring, the inguinal
canal having become dilated, sometimes to very
much greater extent than the internal ring itself,
which may still remain at little more than its normal

size. You frequently find in these cases a large cavity which the hernia has made for itself in the canal between the layers of the abdominal aponeuroses, with no external evidence except a fulness in the groin, and a pain or sense of weakness, increased on coughing or other exertion. At length the pillars of the external ring give way, sometimes by a quick, sometimes by a very slow process ; their edges are first bulged outward ; the external pillar carrying with it the cord, and the internal pillar being relaxed and .curved inwards. This process is effected by the yielding of the arciform bands of the external spermatic or inter-columnar fascia, which gradually allows the pillars to become bulged forwards, and·then everted and curvilinear instead of nearly straight in direction.

Now let us consider how we should best restrain or prevent this process. It must be done by judicious truss pressure, by which even a radical cure may be obtained in some cases. To accomplish this, however, requires so accurate a fit, so appropriate an instrument, and such care on the part of the patient, or, in the case of children, on the part of the nurse, that it is not often successful. About 15 to 20 per cent. of hernia patients may be cured by judicious and persistent truss pressure, perhaps more in the case of children and young persons, and certainly less in adults. The reason why in the herniæ of children we are more likely to get a cure by truss pressure is this :—In children the condition is the result, both in the case of inguinal and umbilical hernia, of an imperfection in evolution shown in the final closing up of the abdominal wall. The former is generally caused by a late descent of the testicle, accompanied, probably, by an imperfect development of the cremaster and gubernaculum testis. The whole of the parts concerned in the descent of the testis are weakened and backward in development ; the consequences do not always show themselves immediately after birth, but may become apparent at some later period of life. Not uncommonly a portion of the peritoneum is drawn down, preceding the testicle itself in its descent into the scrotum. Sometimes the testicle is held in the abdomen by adhesions, while the epididymis is stretched out and unravelled, reaching from the testicle, which remains below the kidney, where it is developed, as far as the external ring.

This fact of the peritoneum preceding the testicle gives rise to a peculiar form of rupture, which you sometimes see in children—viz., the testicle is retained in the abdomen, while a hydrocele or serous effusion forms in the tunica vaginalis, producing a tumour which simulates a bubonocele : a portion of bowel or omentum may now slip beyond the testicle and pass into the sac. The bowel may come down with the fluid in the tunica vaginalis, leaving the testicle behind. These are sometimes called "windy ruptures," and the fluid contained therein may be pressed back into the abdominal cavity. The descent of the testis takes place about the end of the seventh month of intra-uterine life, and ought to be completed before the end of the eighth ; but it varies very much according to the forwardness of the development of the child. The gland ought to be in the scrotum at the time of birth, and it is the duty of the doctor to ascertain whether it is so or not at the birth of a male child. It is generally some weeks afterwards before the canal is closed up. The closure begins at the upper part, and makes its way downward until it seals up the tunica vaginalis about half an inch from the testicle. A similar process, producing a protrusion of the peritoneum into the groin, may occur in a female child, the ovary passing into the labium instead of the pelvis, and sinking further into it as the pelvis develops by an error in development. A process of peritoneum is drawn down with it into the labium, and sometimes this remains patulous, and gives rise to a congenital inguinal hernia in the female. In such a case as that you should apply pressure early, directly backward at the site of the deep or internal ring.

We will now consider the mechanical action of *the truss*. There are two parts of the truss for separate consideration. The chief and most important part is the pad. All trusses for reducible hernia are provided with some sort of pad, whatever may be their principle of mechanical action. The object of this pad is to press upon the opening through which the rupture passes, to keep the bowel or omentum from getting into the canal, and, if possible, to prevent it from slipping down into the scrotum, even if it passes the internal or deep abdominal ring. There are various shapes of pads, some of which are very objectionable in principle. I take one here for example. This is a very conical pad, so conical as to be almost bluntly pointed. Put this on a weak place, and what ensues? There is a hole beneath the integuments, which are spread over that hole, covering it in. The acutely conical truss-pad presses these superficial tissues into the hole, in much the same way as when you put a cork into a

bottle with a piece of leather over it. It thus spreads
the tissues out, stretches and weakens them, and at

FIG. 30.

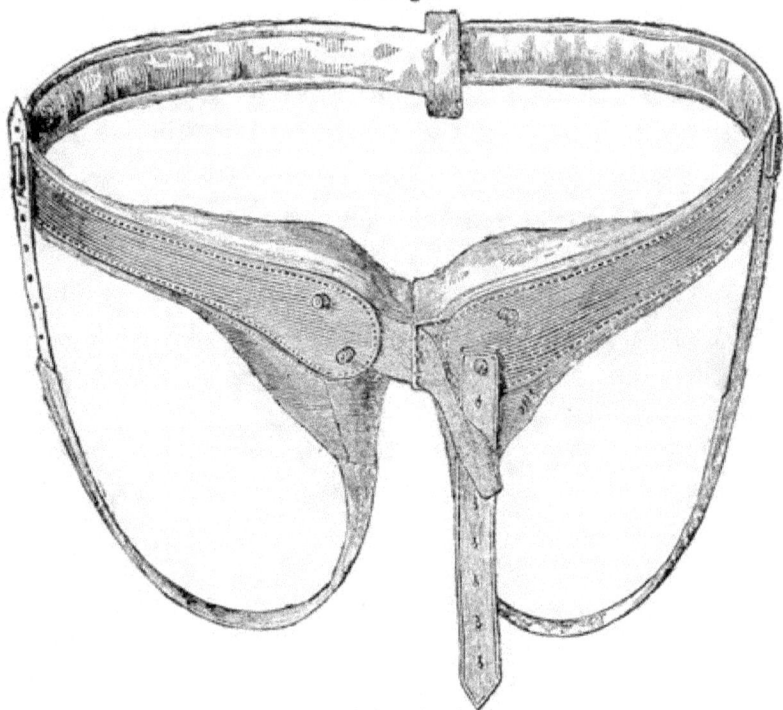

A bad Truss.

the same time dilates the tendinous aperture of the
superficial ring. The
injurious effect of wear-
ing continuously trusses
of this kind, or their
various modifications, all
of which have the same
radical vice, is increased
in some by a powerful
spiral spring placed in-
side the conical pad,
so that they press the

FIG. 31.

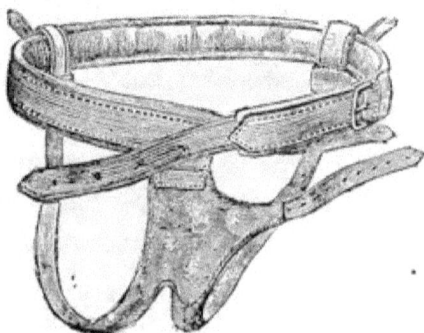

Bad form of Truss.

tissues still more powerfully into the hernial opening. The movements of a patient who wears such a truss causes a constant working of this spring, and a boring motion into the aperture is produced, like the twisting of a cork into the neck of a bottle. Moreover, in a case of rupture you have, not a resisting bottle-neck, but an elastic and valvular opening which yields to the pressure. You are continually obliged, therefore, to increase the size of your cork-like pad, so as to fully occupy the hole and sustain the rupture. The aperture regularly increases, and the rupture when it does come down, constantly becomes larger and more liable to become scrotal, until at length it gets so large and unmanageable that no truss will keep it up. This is a common result of wearing a truss-pad of this kind.

It has been asserted that, as a fact, the projecting conical pad only buries itself in the subcutaneous tissues, and does not project between the pillars of the ring. But this seems a purely arbitrary assumption, and one certainly contrary to the results of observation in cases which have been subject to such pressure for a length of time. If there be no pressure exerted by the point of the cone, what effect can it have in restraining the rupture at all ? If it has any effect at all upon the inguinal canal, that effect must be in accordance with the shape of the impinging surface. A conical wedge-like pressure cannot be transformed into a flat pressure by skin and fat, which are almost as yielding as water. And what occurs in the numerous cases in which the skin is thin and delicate, and the fat almost entirely absent ? A rounded surface must press on all sides in a direction perpendicular to the surface which presses.

It must tend, therefore, to thrust outwards the pillars of the ring, and to stretch and weaken the inter-columnar fascia in the same way as the introduction of a Wutzer's plug into the canal. And that it does so I have verified by post-mortem examination in numerous cases where a truss had been worn. In fact, a conical surface, fitting into a slippery and elastic opening, rather favours the escape of the hernia between its sloping sides and the edges of the aperture, as soon as the rupture acquires sufficient power to lift up the truss-pad a little ; and if it becomes displaced laterally, the hernia immediately slips down the inclined plane. Sometimes in such a case the instrument-maker, finding the rupture to slip down into the canal, and wishing to stop it in its descent, prolongs his truss-pad downwards into a sort of tail, and makes it bigger and bigger, until at length he pushes aside the scrotum, and may bring the pad down into the perineum. Such arrangements hardly ever do any good at all ; they are simply excuses for " fiddle-faddling." They are exceedingly uncomfortable to the patient, and you may be sure, if a rupture gets down the canal so far as to need such a secondary pad, it is certain to pass on into the scrotum ; you cannot stop it. The most important indication, therefore, is to prevent the rupture from entering the canal at all ; to shut up the internal ring altogether.

The pads are fastened on to the retaining appa-ratus in a variety of ways. Some are made so as to be adjusting, with the idea of following the rupture in the various twistings of its course and emergence. After long experience, I have come to the conclusion .that nothing useful can be done by such pads as these, unless the patient is constantly on the watch to adapt the pad to the shiftings of the rupture. This, I need not remark, can scarcely be done in society, or in the streets, or in various situations where the stress of a rupture may come. The best way is to have your pad so fixed that it prevents the rupture from getting into the canal at all. The kind of pad I myself recommend—what I chiefly insist upon—is, first, that the bearing of the surface of the pad shall be *flat*, that it shall not press in the tissues or invaginate them into the canal between the pillars of the external abdominal ring, and thus stretch, fray, and weaken the inter-columnar fascia, which ought to restrain the rupture from coming down. The edge of course must be rounded off to prevent it cutting. We get, therefore, to this kind of flat-bottomed-boat shape, according in outline with that of the inguinal canal in its diseased condition; that shape is an oblique oval. In oblique inguinal hernia in the female a flat oval pad, without any break in its outline, answers very well indeed. In the male, however, there is a peculiarity in the anatomical arrangement of the parts. The spermatic cord passes out of the superficial abdominal opening external to the spine of the pubis, crossing or lying over the outer pillar of the ring. If the truss-pad produces a pressure upon the cord, it not only makes the patient uncomfortable and the pad more liable to shift and slip about, but also may cause swelling of the testicle, hydrocele, varicocele, and ultimately atrophy; while the chafing of the pad against the pubis leads to the formation of excoriations, sores, and even abscesses. In order to avoid this, we have in the truss now before us a

chink or slit in the pad (see figs. 32 and 33).   This
gives the pad a sort of oblique horseshoe shape.   If
properly put on, the shorter end lies upon Poupart's

FIG. 32.

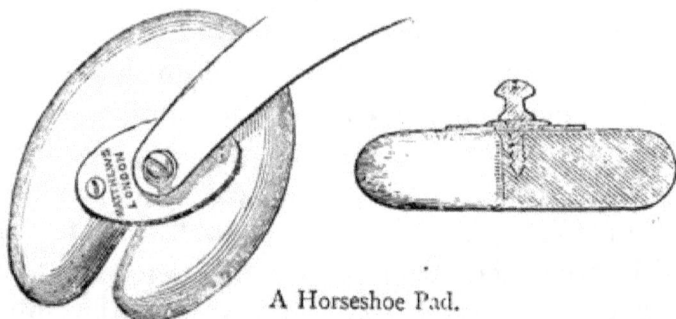

A Horseshoe Pad.

ligament immediately outside and above the spine
of the pubis ; the longer end lies on the inner pillar,
and the round end covers the deep hernial opening
of the external ring.   The cord, being very movable,

FIG. 33.

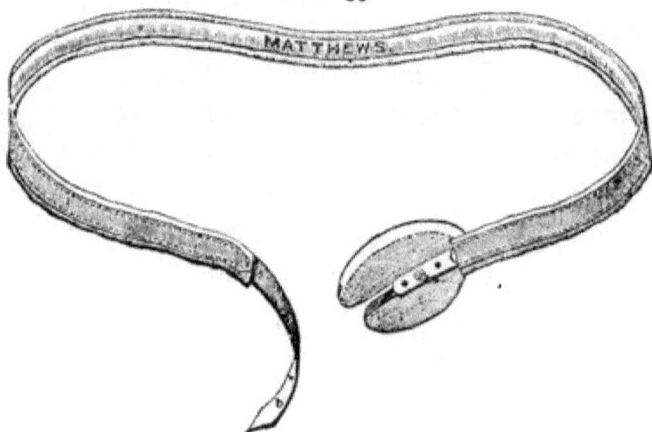

A Truss with Horseshoe Pad.

will adjust itself to the pad and slip into the part
where there is the least pressure,—i.e., into the chink
left between the two points of the pad.

If you hope to get the inguinal canal closed up, and the sac obliterated by a radical cure, it is better to have the pressure hard and firm ; and the best material for the pad is a substance called "vulcanite," of which a specimen is before you. This does not absorb the perspiration, is perfectly smooth and hard, and, if proper care is taken to keep the surface clean and dry, will not chafe or give rise to sores. The next best substances are boxwood and ivory, which, however, absorb the perspiration to a somewhat greater extent. Experience of these trusses in hot

FIG. 34.

Plated Bathing Truss with Vulcanite Pad.

climates has been unanimously in favour of the hard vulcanite over any other substance for truss-pads. Leathern or parchment coverings become putrid, foul, and hard, under the effect of constant absorption of the cutaneous excretions, and get so nasty that sensitive and cleanly patients cannot bear to wear them. In other cases, where you do not try for a radical cure so much as for making a patient comfortable, then water and air pads, made up of india-rubber upon a metal frame, are exceedingly

comfortable and useful.   Some patients cannot bear
any other than this soft pressure.   They are of the
same general shape and principle, but the surface is
more yielding, and the pressure is soft ; they cannot
press into the hernial apertures as hard conical pads
do, and the pressure, being fluid, is equal in all
directions.

The nature of the apparatus for fixing and keeping on the pad and restraining the rupture is also of
great importance.   The rupture requires pressure to
retain it, and, as a rule, you do no good at all unless
there is a side-spring.   There have been various
ways devised of applying this retaining apparatus.
One way which the patients sometimes choose, and
which seems to recommend itself to them by its
simplicity, is having a strap round the body, and an
understrap across the perineum.   Now, it is exceedingly difficult to wear a band round the waist so
tight as not to give way to pressure at one point,
and so to yield before the rupture.   Even if you
could make it tight enough, the patient could not
wear it, the constriction would be so great.   You may
take it as a rule that these straps round the pelvis,
when a patient is exerting himself and contracting
his abdominal muscles, are of no use in keeping in a
rupture.   Where there is real need for pressure,
nearly all truss-makers have recourse to some form
of the side-spring.   Some have the spring passing
only round one-half of the body, with a pad behind
on the sacrum.   This pad is always flat or oval, and
slightly concave, and is larger and thinner than that
placed on the rupture.   It is held in its place by a
strap that goes round the opposite side of the body,
and frequently by an under-strap across the perineum.

Some have the spring put on the same side as the rupture ; but in Salmon and Ody's truss (fig. 35) the spring is put on the opposite side, so that it reaches across the front of the abdomen, and is longer than those which are put on the same side as the rupture. The rupture pad projects more than the posterior one, and works upon a ball-and-socket joint. The spring is longer than is necessary to go only half round the body; it reaches over to the opposite side,

FIG. 35.

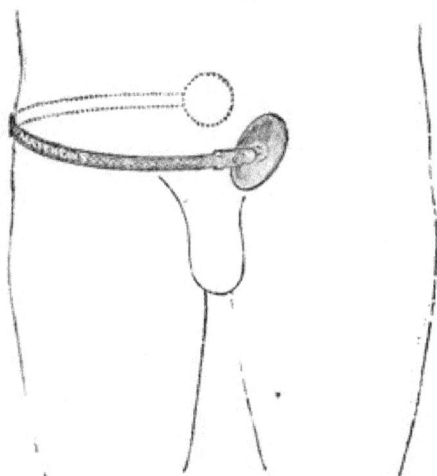

Salmon and Ody's Truss.   Spring passes over pubes and presses on reverse side.

and the support it gives depends upon the fact of its pushing upwards and towards the ruptured side. Other half-round springs depend upon the power they possess of pressing or pulling upwards and outwards towards the same side, in opposition to the descent of the rupture.   Sometimes it is necessary to wear a perineal band which buttons in front. Generally speaking, this may be dispensed with after the truss has accommodated itself to the shape of the body, which after a time all trusses do to some extent. The warmth and motion of the body will make even the spring accommodate itself somewhat to the shape of the body.

With regard to the bend of the spring, there are one or two matters of very great importance. In the first place, the spring should go round the body at a level midway between the projection of the tro-chanter and the anterior superior iliac spine. There it lies on the tensor vaginæ femoris and gluteal muscles, and does not work over bony surfaces. That is the level at which the measurement for a strap should be taken when you have to send for a truss to a maker. But I may here remark that it is never satisfactory to send measurements without the maker seeing the patient. It is as if you were to send measurements to a tailor or a shoemaker to make your trousers and shoes from. You would scarcely be likely to have a good fit. The maker wants, in addition; to comprehend the shape of the back and set-on of the pelvis as well as the mere dimensions. So that, if possible, you ought to bring the maker and patient together. Other plans do not usually succeed. The spring should point down far enough to get to the opening, and the pad should be placed upon the opening. You will see that the spring requires to be bent down a good deal more for crural than for inguinal hernia. For crural hernia the side-spring should be made like the handle of an old-fashioned pistol. In the inguinal variety, that end of the spring which bears the pad should project well, so as to give a proper degree of backward pressure ; and if you look at the surface of the spring you will see that it is somewhat twisted on its own axis, so as to give an outward and upward pressure as well as a backward pressure. This gives the right direction in which to keep the hernia in the abdomen when it tends to pass into the canal. The round part of the horseshoe pad presses upon the internal ring, and the ends press upon the pillar of the external ring. The chink is to lodge the cord, which is thus held as if embraced by the fingers employed in reducing and keeping in the rupture. Thus the rupture is prevented from coming through the internal ring, while the pillars of the superficial opening are prevented from separating, and so allowing the rupture to pass out. The length of the spring from the point where it comes round the hips should be duly proportioned to the patient's formation. In these horseshoe pads there are holes and screws by which the pad can be shifted a little to adjust this properly. If the spring be too long at this part the pad presses against the outer edge of the rectus muscle. The inner border of the pad should be parallel to the outer border of this muscle, and the outer border should lie upon Poupart's ligament. If the spring be too long it pushes the pad further on to the muscle, which bears off the pressure from the hernial cause during the contraction of the muscle, permitting the rupture to escape below and outside the pad. If, on the other hand, you have the spring too short, the rupture will escape between the rectus and the pad.

A *direct inguinal hernia* passes through the triangle of Hesselbach, enclosed between the epigastric artery, the edge of the rectus, and Poupart's ligament. That is the area you have to protect ; and it can best be done by a flat-rounded or oblately-oval pad (figs.

36 and 37) fitting close between the edges of the rectus and Poupart's ligament, reaching well down to the crest of the pubis, and provided with a slight notch below for the passage of the cord.   To keep the pad

FIG. 36.

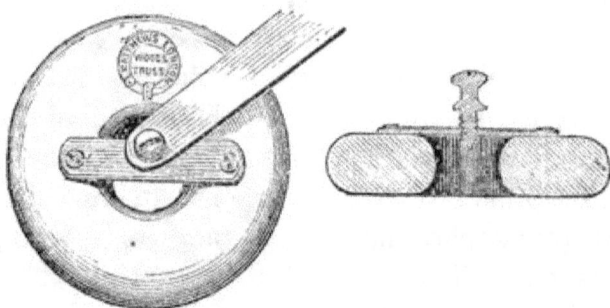

from shifting upwards and from pressing unduly upon the pubis, care is required in adjusting the action of the side-spring.   It is as well to wear at first an under or perineal strap, until the pad and

FIG. 37.

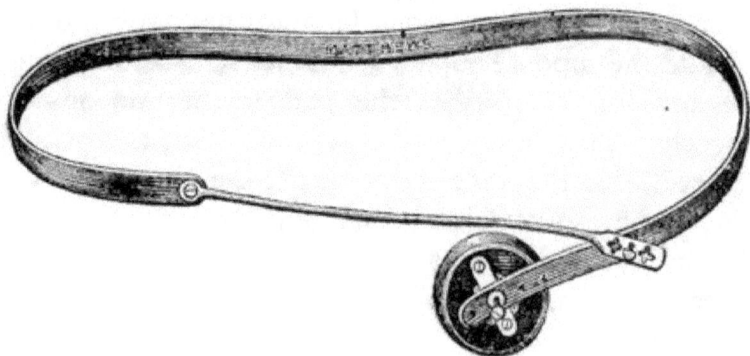

spring have adjusted themselves to the shape of the abdomen.   In corpulent persons a considerable upward slope must be also given to the surface of the pad, to make it lie parallel with the slope of the

abdomen, and to prevent the upper edge from pressing unduly into the flesh.   In thin persons, with lean
flanks, the tendency is always for the pad to slide
upwards into the hollow formed by the abdomen.
This can be met sometimes by keeping the surface
of the pad quite flat, so as to lie perfectly level upon
the surface of the groin.   You may, however, in the
course of time, in the same patient, find a marked
alteration in the slope of the abdomen from an
increase in the abdominal volume as well as in the
thickness of the superficial fat, altering entirely the
conditions of the rupture and the requirements of
the pad and spring.   In some instances this may
occur in a very short time.   On the other hand, a
patient, from illness or active work, may get rapidly
thin, and require a readjustment from this cause.
To meet and manage these conditions is one of the
niceties of truss-making.   It is sometimes difficult
to get the exact twist, and even when you have got
it right the condition of the patient may change, and
you may have to alter the spring accordingly.

The problem to solve may be put geometrically :
It is requisite to obtain the angle of inclination of
the abdomen to a transverse vertical plane, taken at
the most prominent part of the inguinal region, and
containing the side of a right-angled triangle, of
which the posterior wall of the inguinal canal is the
hypothenuse, and the horizontal level of the upper
margin of the pubis is the base.   If you do not have
the pad-surface inclined enough, the rupture comes
down under its lower border ; and if you twist it too
much you get the same edge pressing in so as to
inconvenience the patient, and allow the rupture to
enter the upper part of the canal.   It is this slipping

I

over and under on one side or the other that constitutes the troublesome part of the treatment of rupture by trusses. You do not often find patients who have sufficient mechanical knowledge, or who take sufficient pains, to aid the efforts of the instrument-maker by skilful adjustment of the pad after a careful return of the rupture. This is one cause why so few cures are effected in this way. I recommend, as a rule, the all-round spring covering over both hips instead of the one-sided spring ; but in certain cases I believe the principle followed in Salmon and Ody's plan, a half-round spring, fitted to the opposite side of the hips, and pushing towards the ruptured part, may be advantageous ; the horseshoe form of pad, however, may be used quite as well with this form of spring (fig. 35).

In old cases of irreducible hernia you meet with another difficulty—you cannot reduce the hernia entirely, and all you can do is to prevent more of the intestines from coming down. In such cases the bowel may be exposed to all sorts of injury, besides constituting a deformity of a somewhat conspicuous character. To support this you must have a suspensory or bag-truss made of stout jean, or some unyielding material, which will keep a constant pressure upon the contents. If the irreducible portion consist of omentum only, you must also have some pressure over the inguinal canal to prevent the bowel from following the omentum. Such combinations are sometimes exceedingly difficult to carry out. Messrs. Matthews have tried, in some of my cases of this kind, with much success, a truss-pad shaped to the form of the rupture, composed of a frame of stout wire, well padded, and stretching between the wire framework a bag of stout jean, or of some slightly elastic material, sufficiently resisting, into which the hernia is received (fig. 39). The wire framework, pressing all round the irreducible rupture, keeps it well in hand and under control. All you can do in such cases is simply to make your patient as comfortable as circumstances will allow, and to prevent injury to the irreducible rupture.

FIG. 3S.

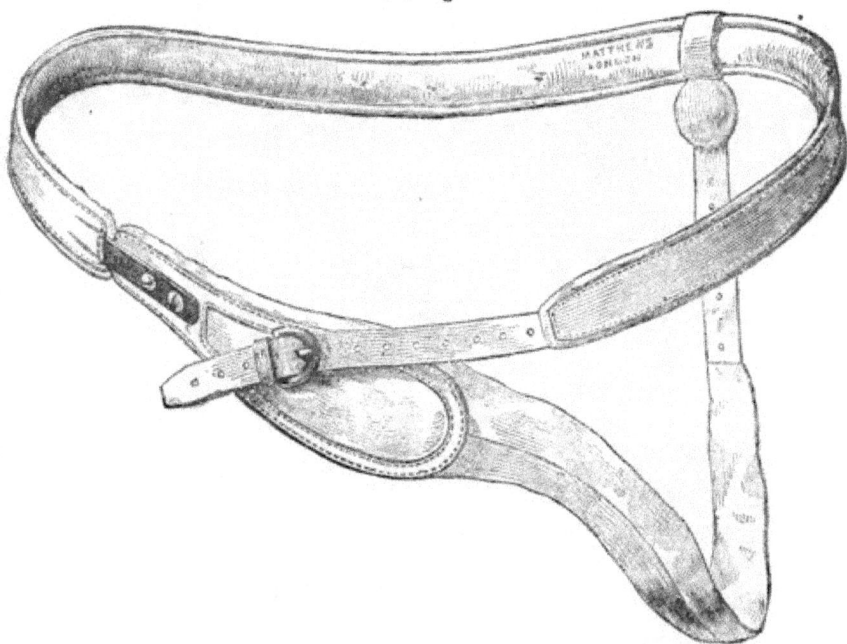

Drumhead Pad for partially reducible Hernia.

In *crural hernia* we have conditions entirely different. The inner opening is constituted by a crural ring, a horizontal aperture with a slight inclination forward. In front it is bounded by Poupart's ligament, on the inner side by Gimbernat's ligament, and on the outer side by the femoral vein and artery.

These are structures which vary somewhat in tension. Relaxation of the muscles of the abdomen has a great effect upon Poupart's ligament. But the greater part of the surrounding structures are com-

FIG. 39.

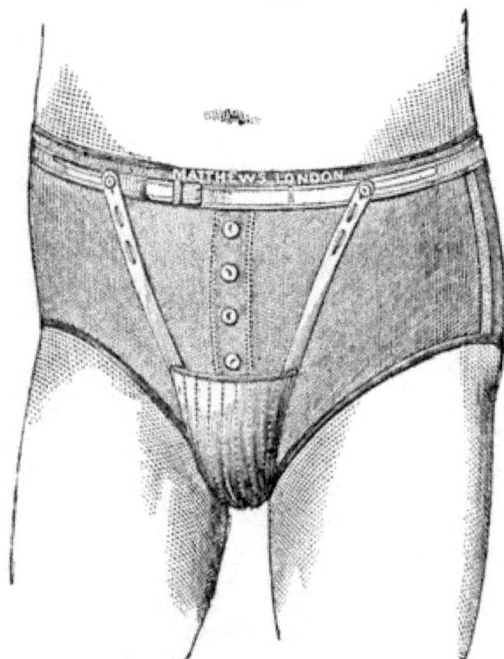

An elastic silk Belt, the fibres of which stretch *vertically only*, to give a supporting pressure to the hernia, supplemented when required by an elastic strap to give additional support to the scrotum. It is usually worn over the shirt, and is recommended for old people and irreducible cases of long standing.

posed of unyielding ligamentous tissue, so that there is not that contraction and relaxation that is present in inguinal hernia. A little below there is another opening, called the saphenous opening, directed forwards and a little inwards, and almost vertically, but with a slight inclination downward. Lying in front of the passage between these two openings is the upper part of the process of Burn's or femoral ligament (Hey's), which extends from half to three-quarters of an inch downwards from Poupart's ligament, with which it is continuous above, to the margin of the saphenous opening.

It is this part to which the pressure of a truss should be applied in crural rupture, when it will protect both crural ring or upper and the saphenous opening or lower aperture of the crural canal. Immediately outside the canal are the femoral vein and artery, which must not be pressed on by the truss ; and below is the saphena vein, which it is also important not to compress.

When a femoral rupture gets fairly through the saphenous opening, it turns upward and outward round the edge of the falciform process, and lies over the femoral vessels and upon Poupart's ligament. In order effectively to deal with this rupture, you must altogether prevent it coming through the crural ring into the canal before it makes the upward and outward turn, so as to lie upon Poupart's ligament. If you fail in this, then your truss, pressing the rupture against the falciform process of Burns, thereby injures the bowel, and does harm rather than good, and the patient would be safer and better without a truss at all.

The truss-pad for crural hernia must protect the crural ring by pressure over Poupart's ligament, and it must also press upon and fill the saphenous opening. It must not press downward, so as to obstruct the saphenous vein. The pad will be apt to slip, so as to miss the crural canal altogether, and, by irritating the inguinal glands, may cause trouble. The best form of truss-pad for this hernia is the one I show you (fig. 40). The outline is an egg-shape,

with the small end downwards : it is adapted to the
saphenous opening, but rather longer, so as to press
upon Poupart's ligament with its broad end above.
The side-spring is fixed exactly in the centre. If
you look at the section, you will see it slopes off
below, so as to avoid pressing upon the saphenous
vein, and forms a rounded projection above, so as
to fall into the fold of the groin upon Poupart's
ligament when the patient sits down. It is thus
adapted for keeping in position ; for the truss-pad
which most adapts itself to the form of the surface
will stop in its place the best.

FIG. 40.

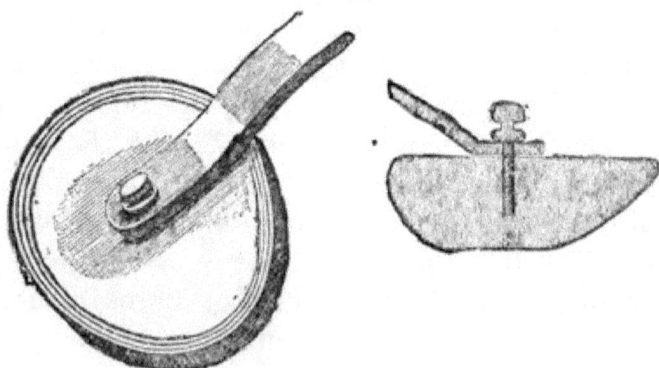

Femoral Pad.

In the truss for femoral hernia, the pad end of the
spring is bent downwards in a large curve to permit
the patient's thigh to bend freely and without
obstruction in sitting. You ought not to be content
with seeing your patient stand when you fit on a
truss ; you must make him sit down on a low seat,
and then stand, walk about, and jump from a stool,
and see if that dislodges the truss. If the truss
does not hurt him, but keeps the hernia up, under

those conditions, you may conclude it will do for all
the ordinary purposes of life. The commencement
of a radical cure by truss pressure always dates from
the last time the bowel or omentum came into the
sac of a rupture. Hence the importance of the
patient preventing the hernia from ever coming
down. If it come down even once, he has to begin
de novo from that point to produce the obliteration
of the canal. Hence a patient who wishes to get
rid, by truss pressure at the earliest possible period,
of a troublesome deformity, must wear his truss on
all occasions, night and day ; he must never assume
the erect posture without it ; and if he bathes, he
must have a bathing truss, for sometimes in the
gymnastic movements which generally attend upon
a cold bath the rupture may come down.

One or two words with regard to *umbilical rupture*
and its apparatus. Umbilical rupture is exceedingly
common in children, and in them it is usually curable.
It comes through a natural opening which is left for
the umbilical vessels up to the time of birth, and
which it is the tendency of nature to close up sooner
or later. That tendency is very strong, and the
only thing that prevents it is the bowel constantly
coming into the sac.. If, in a child, you can manage
to prevent this, you cure the hernia ; and that is
generally the case when the improved apparatus of
the present day is carefully attended to. But there
are some cases where the child is not tractable, and
from pain and fretfulness is often crying and scream-
ing ; then you get the rupture distended violently
and constantly. Again, if the nurse is not skilled
and careful, you seldom get the rupture cured. The
ordinary rough-and-ready and often very successful

fashion of treating umbilical hernia is covering a flat
piece of metal—say one of the bronze coins of the
realm, a penny-piece—with adhesive plaster, with
the sticky side outwards, putting it on the projection,
and strapping it across and nearly round the abdomen
with broad straps.

In some of the older books on this subject you
will find recommended a convex cork plugging up
the aperture, like the neck of a bottle ; but elastic
apertures of vital tissue cannot be blocked up in that
way, while the cork tends to make matters worse by
dilating the aperture, and thus keeping open the
rupture.   Therefore that is one of the things to be
avoided.   It does not keep in the rupture, because
this will slip out at the side of the cork.   A flat
surface, rather larger than the aperture, is what you
ought to have.   A flat penny-piece, or bit of lead of
the same shape and size, may be backed up by a
thicker piece of wood or cork, and the strapping
may be put across.   In this way a very good and
easy appliance is made, if the patient cannot afford
to have a proper apparatus; but it involves the
necessity of a tedious process of taking off sticking
plaster, which is sometimes not done in the gentlest
way, and thus sets up a crying bout, which brings the
bowel out of the aperture.   All this is inconvenient.
When you are called upon to do this, you must press
the parietes of the abdomen firmly together with your
finger and thumb, so as to close the umbilical hole
before you take off the pad and strapping, and take
care the bowel does not slip out.

A capital invention is one produced by Messrs.
Matthews, a very ingenious adaptation of elastic
india-rubber, arranged in two compartments, dis-

tended with air, and communicating by a small
aperture: a central one, globular in shape, and an
outer ring. The former presses upon the umbilical

FIG. 41.

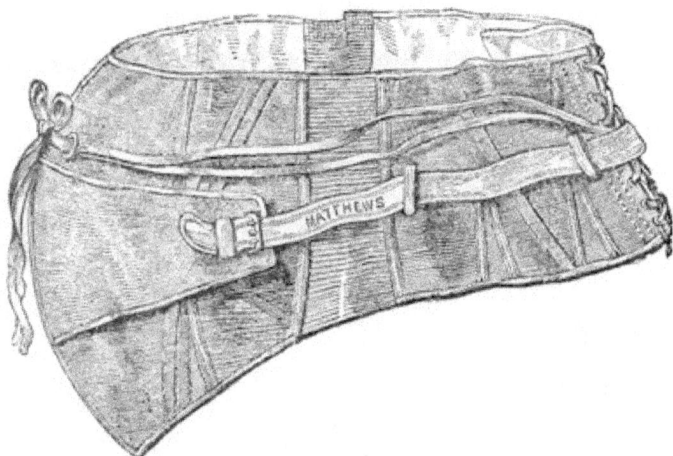

Abdominal Belt, which may be fitted with reflex or other pad it
required.

opening, and the outer upon neighbouring tissues
forming its boundaries, and so prevents the umbilical hernia from coming out under a cough or cry

FIG. 42.

Reflex Pad.

impulse (figs. 41 and 42). This central portion
acts like the penny-piece, with the additional advantage of becoming tightly distended by the air

K

from the surrounding ring cushion, forced through
the small aperture of communication by the impulse
of the abdominal muscles.   The whole is held on by
an elastic band round the body, and can be distended
after fixing by blowing through a little stop-tap.   By
this means the moment the bowel has a tendency to
escape through the hernial aperture, it is met and

FIG. 43.

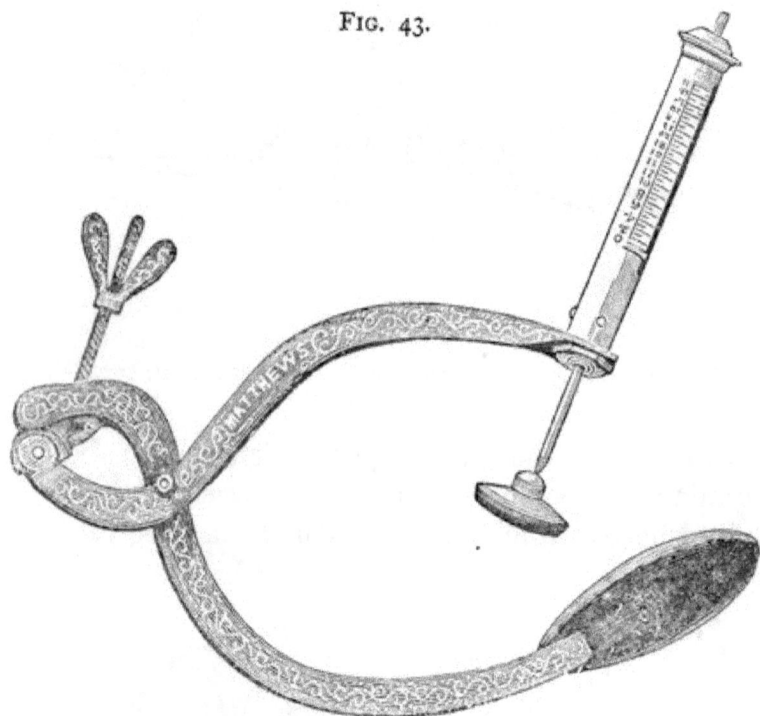

Wood's Pressure Gauge for ascertaining the amount of Hernial Impulse
and consequent strength required for Truss Spring. (Vide *British
Medical Journal*, Oct. 14, 1871.)

forced back again by the dilating globe.   By this
means have been produced some very capital cures in
children.   In adults also, in whom it is much more
difficult to produce a radical cure of umbilical hernia,

this apparatus is very useful and comfortable. The chief reason why in adults you do not get a radical cure of umbilical hernia is because it is generally accompanied by abdominal obesity and laxity, the stomach too becoming at intervals much distended with food and flatulence, and the mesentery being enlarged by an accumulation of fat. In such persons you must be content, even in any kind of hernia, with amelioration of their condition rather than cure. But in young persons you may frequently succeed in effecting a cure by the aid of a proper instrument. If, in young persons a hernia of the inguinal or umbilical variety resists the cure by careful mechanical restraint, then it becomes a question whether you cannot safely and greatly increase the chance of a cure by an operation, which keeps out the bowel for a sufficient length of time for the opening to contract and close.

FIG. 44.

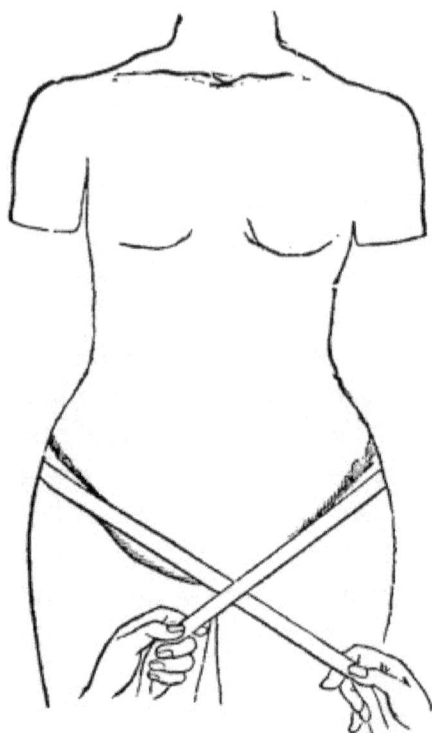

MEASUREMENTS AND PARTICULARS REQUIRED IN ORDERING A TRUSS.

State whether the hernia or weakness is on the right, left, or on both sides.

Give some idea of the size of the protrusion—such as, large as a walnut, egg, &c.

State whether the opening through which the hernia escapes is large or small.

State description of hernia.

Measurement for inguinal hernia (state whether oblique or direct), the girth of body half-way between the iliac crests and great trochanter, —the tape meeting in front. In extremely bad cases of hernia it is sometimes necessary to take a plaster of Paris cast of the pelvis.

BALLANTYNE PRESS, CHANDOS STREET, W.C.